LISA

Molecular Neuroscience

Mon 18th 9.30 - 12.30 elements 1+2

Fri 22nd 9.30 - 12.30 " " 3+4

tutorial rm 2.9

teaching labs

→ Prof Mitchells office E27 @ 9.15

Cell and Molecular Biology in Action

Edited by: Dr Ed Wood, Department of Biochemistry and Molecular Biology, University of Leeds, UK

The series aims to provide introductions to key, exciting areas of cell and molecular biology, stimulating students' imaginations and initiative to bridge the gap between memorising concepts and the active approach needed for research and literature review projects. This active learning series also introduces students to experimental design and information retrieval and analysis, including exploration of the World Wide Web.

Each text in the series will cover key theory concisely and use boxes to highlight skills, techniques and applications of the theory covered. Each text will also have its own Web page providing updates and useful links to relevant sites.

For details of forthcoming titles in the series please visit the Addison Wesley Longman World Wide Web site at http://www.awl-he.com/

CELL AND MOLECULAR BIOLOGY IN ACTION SERIES

Molecular Neuroscience

David Carter and
David Murphy

Longman

Pearson Education Limited
Edinburgh Gate, Harlow
Essex CM20 2JE
England

and Associated Companies throughout the World

First published 1999

ISBN 0 582 35702 0

British Library Cataloguing-in-Publication Data
A catalogue record for this book is available from the British Library.

Set by 30 in Concorde BE
Printed in Great Britain by Henry Ling Ltd., at the Dorset Press, Dorchester, Dorset.

Contents

Further information for use with this book can be accessed via a link in the catalogue entry for *Molecular Neuroscience* on the publisher's Web site at **http://www.awl-he.com/biology**. The aim of this facility is to provide updates and news of major advances in molecular neuroscience since publication. It also includes study guides for chapters, and cites additional Web site addresses which provide information of interest to molecular neuroscience. We would encourage readers to use the Web site regularly.

Preface

This book has been developed from final year undergraduate and postgraduate courses in molecular neuroscience which the authors have devised and taught. The text is therefore aimed primarily at advanced university students, but we hope that it will also be useful for research workers; both neuroscientists who are adopting molecular approaches for the first time, and also those already in the field.

The book is intended to introduce the field of molecular neuroscience, a new and dynamic research area which is seeking to understand the molecular mechanisms of complex brain functions such as memory, and the molecular pathology of inherited neurological diseases such as Alzheimer's disease. It is hoped that this text will help to bridge the gap between the basic understanding which can be gained from more general texts, and the 'cutting edge' of a particular research area. In order to achieve this, we have considered it necessary to include a significant amount of molecular biology in the early chapters of the book. Therefore, 'molecular-literate' students may wish to focus initially on Chapters 6–10, which deal specifically with molecular aspects of neuroscience. However, the core material in Chapters 1–5 is explained with reference to many recent developments in molecular neuroscience, and these are linked to neurophysiological and neuropathological topics described in the later chapters.

The text is divided into four sections which describe the identification, manipulation and regulation of neuronal genes, and also the role of neuronal genes in health and disease. Throughout the book, journal references are introduced which will guide the reader into a deeper appreciation of a particular topic. A key aspect of the text, however, is the reference to Web sites – of leading individual researchers, discussion groups, biotechnology companies and more. The growing importance of these sites cannot be over-emphasised. The pace of research in molecular neuroscience is such that printed journal articles become rapidly outdated – electronic publishing is the medium of the future, and our text has been designed to interface with a Web site that will enable students to begin searching this vital resource. Students should note, however, that much of the material published on the Web is not subject to the rigorous peer-review process which maintains the high standards of conventional research publications, and therefore care should be exercised in the interpretation of some material.

We would like to express our thanks to research colleagues, collaborators, friends and family who have contributed to, and supported our work, and therefore made possible the writing of this book. In particular we would like to thank Professors Vincenzo Crunelli, Tony Cryer and Stafford Lightman for support. We are also very grateful to our 'guests' who have made valuable contributions to the book. Finally, we were ably guided through the writing process by Alex Seabrook, Kate Henderson and Lynn Brandon of Pearson Education, and we would like to thank them for their enthusiastic support.

CHAPTER ONE

Cloning and analysis of neuronal genes

Key topics

- cDNA cloning
  - cDNA library screening
  - PCR-based cDNA cloning
  - Sequence analysis of cDNA clones
  - *In silico* cloning
  - Applications of cDNA clones
- Gene families and orphans
- Alternative splicing
- RNA editing
- ESTs
- Genomic DNA cloning
  - Genomic libraries
  - Cosmids, BACs and YACs
  - PCR-based genomic cloning
  - Applications of genomic clones

1.1 Introduction

Molecular neuroscience is a new scientific discipline which represents an increasingly large proportion of contemporary brain research. The reason for the burgeoning of molecular studies is clear - despite continued, and important, advances in the classical techniques of electrophysiology and neuroanatomy, the power of new molecular genetic techniques has enabled neuroscientists to make rapid and incisive advances in our understanding of the brain. There are two major aspects of these recent molecular studies:

- **Gene cloning** – knowledge of the DNA sequence of brain genes has provided precise information about the molecules which 'drive' the brain.
- **Genome manipulation (transgenesis)** – the ability to manipulate the mammalian genome in a DNA sequence-specific manner has enabled precise functional studies of brain genes.

Both of these aspects will be described in this book. Although readers may be familiar with some of the basic molecular biology covered in the early part of the book, this material has been selected in order to clarify more complex issues described in the later chapters. Also, the basic molecular biology is introduced with reference to important new discoveries in molecular neuroscience which show that our understanding of basic molecular concepts is also being informed by research in this field. Molecular neuroscience is therefore more than just the application of molecular techniques to the study of the brain – it is now a fully integrated aspect of modern biology.

The rapid, world-wide progress in molecular neuroscience is staggering and it is difficult, even for full-time researchers, to keep up with new developments. This book should be seen as a bridge between the major questions of molecular neuroscience, and current work in the field. Through reference to Web sites (which can provide up-to-the-minute research data) and other source material, it is hoped that the reader will be able quickly to grasp the major aspects of a particular research area, and then apply their own imagination and initiative to the current problems which lie at the heart of the research. The use of Web sites is described in Box. 1.1

BOX 1.1: WEB RESOURCES

The World Wide Web is a simple to use graphical interface that gives access to information stored on the computer Internet. In order to get started on 'the Web' and use the Web resources listed in the book, it is assumed that students will have access to an Internet-linked personal computer, either in college or at home. A familiarity with Web 'browsers' such as Netscape is also assumed. Novices should note that by virtue of the 'point and click'-based operating systems, much of Web 'surfing' is self-evident and little or no instruction is required.

The Web site for this book can be accessed via a link in the catalogue entry for *Molecular Neuroscience* on the publisher's Web site at http://www.awl-he.com/biology. From this site, access can be made to other sites listed in the book. A single important site for Neuroscientists is that of *Neuroscience on the Internet* **(http://www.genetics.gla.ac.uk/neil/index.html)**, which contains multiple links to other sites, and therefore sufficient information to fill most average lifetimes!

A single useful reference is Bloom (1996) which, although somewhat dated now, provides the considered view of a distinguished neuroscientist on browsing the Web for neuroscience information. Finally, the caveat mentioned in the preface to this book should be re-stated: much of the material published on the Web is not subject to the rigorous peer-review process which maintains the high standards of conventional research publications, and therefore care should be exercised in the interpretation of some material.

The pioneers of molecular genetics first started to clone genes in the mid to late 1970s (see Hall, 1987 in Further Reading), and since then the generation of novel gene sequences has increased year by year. In the field of neuroscience, molecular biologists have now sequenced thousands of molecules which are crucial for brain function, including neurotransmitter receptors and ion channels (see Green *et al.*, 1998). However, a consideration of the number of genes expressed in the brain (50 000 of the total 70 000 human genes) caused certain research groups to adopt a more global and opportunistic approach (see Milner and Sutcliffe, 1983) because it was apparent that a proportion of the unknown genes would almost certainly code for novel molecules with important functional and disease associations. This idea has contributed to the current genome-wide DNA sequencing projects which are discussed in Chapter 2.

1.2 cDNA cloning

Molecular cloning [cf. cell or organism (e.g. sheep) cloning] can be defined as the incorporation of a gene sequence into a **vector** that may be propagated to produce further copies of the sequence. cDNA cloning involves the manufacture of **complementary (c)DNA copies** of genes because although the coding sequences of genes are contained within messenger (m)RNAs (Figure 1.1), the utility of RNA molecules is limited by their exceptional lability.

1.2.1 cDNA libraries

A cDNA is a synthetic molecule which does not occur naturally. It is synthesised in the laboratory using an enzyme, **reverse transcriptase**, which acts upon mRNA templates to produce reverse complement DNA sequences (Box 1.2). The starting point of cDNA cloning is therefore mRNA which can be readily extracted from tissues (e.g. whole brain or, alternatively, dissected brain regions).

For most purposes, cDNA libraries are used in preference to genomic libraries (see section 1.3) because the protein-coding sequences are, at least, the initial goal of the research and also cDNAs are conveniently small relative to genes (which can contain very large introns; Figure 1.1).

The technique of cDNA library construction (Box 1.2) has been refined over the past 10–15 years, and has become highly efficient. However, it is now possible to purchase a wide range of species- and tissue-specific cDNA libraries from molecular biology companies (e.g. Stratagene catalogue: **http://www.stratagene.com**), a short-cut which can both save time and eliminate potential biohazards including the handling of human brain tissue. Such companies are also actively involved in improving technologies, and have become an integral part of the research community. A recent innovation at Stratagene, for example, has highlighted the use of high-quality plasmid cDNA libraries which can often replace the less convenient bacteriophage-based libraries:

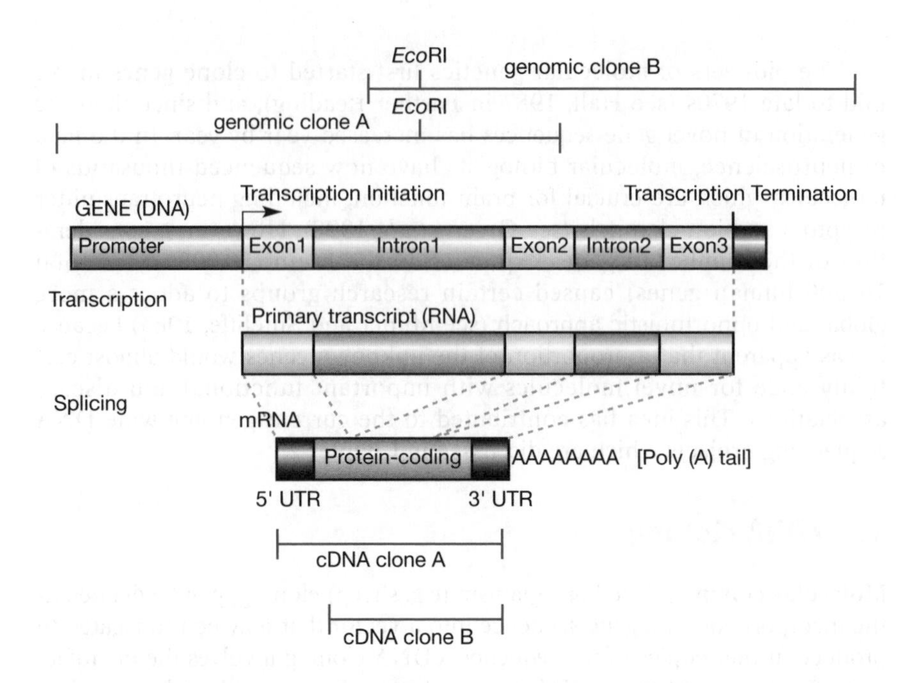

Figure 1.1 Eukaryotic gene organisation and derivation of primary (precursor) RNA and messenger (m) RNA. Introns may cover extensive regions of the gene, separating exons with many kilobases of intronic sequence. In addition to protein-coding sequence, the mRNA includes both 5' and 3' untranslated regions (UTR) which act in the regulation of translation and mRNA stability.

Note:
1. Genomic clones (fragments) may not represent the entire gene sequence. In this example, genomic fragments A and B may be cut and joined (ligated) at the *Eco*RI restriction enzyme site, forming a new fragment which represents the entire coding regions of the gene.
2. A full-length cDNA corresponds to the entire mRNA sequence (e.g. cDNA clone A). Clones lacking 5' sequence (eg. cDNA clone B) are often obtained, and additional procedures must be used to obtain the missing sequence (see section 1.2.2).

BOX 1.2: cDNA CLONING

A cDNA library is a collection of cloned double-stranded DNA copies representative of the mRNA pool of a tissue sample. A short double-stranded region is generated by the annealing of a primer to the mRNA. There are several options for first-strand primers – a specific sequence complementary to the target mRNA, an oligo dT primer that will anneal to the poly(A) tail at the 3′ end of mRNA, or short random oligonucleotides that anneal to many internal locations along the length of the target RNA. The DNA strand complementary to the RNA ('first strand') is then generated by the action of avian myeloblastosis virus (AMV) or Moloney murine leukaemia virus (MoMuLV) reverse transcriptase. Older methodologies took

BOX 1.2: Contd

advantage of the fact that this first-strand synthesis results in the production of a short hairpin loop that provided the primer for the second-strand synthesis reaction. Subsequent S1 nuclease digestion was then required to remove this single-stranded hairpin region. This approach has been superseded by a method that uses DNA polymerase I in combination with RNase H to replace the mRNA in the hybrid with small regions of newly synthesised DNA. These short stretches of double-stranded DNA are ligated into a continuous strand by T4 DNA ligase. At the same time, linkers containing a restriction endonuclease site, for example, *Eco*RI , are ligated to the blunt ends of the cDNA and then digested with the restriction enzyme to allow cloning into any similarly cut vector. *Eco*RI sites within the cDNA itself can be made resistant to cleavage by prior methylation. The representation of a particular sequence within a cDNA library is a measure of its expression in a tissue, and sequencing of cDNA clones can provide information about the structure of the mRNA, especially when compared with the sequence of the corresponding gene.

http://www.stratagene.com/vol10_3/p119-120.htm

Although commercially available cDNA libraries (see section 1.2) will serve the majority of purposes, it is important to remember that construction of a customised library may be the route to a research breakthrough. For example, cloning of the melatonin receptor proved elusive for many years until an **expression library** (see section 1.2.2) was constructed from toad (*Xenopus laevis*) dermal melanophores – a rich source of this particular receptor (Ebisawa *et al.*, 1994).

cDNA libraries are representative of a multitude of different genes, and therefore a screening procedure is employed to select the clone of interest. The preparation of cDNA libraries for screening is described in laboratory manuals such as the famous 'Maniatis' (Sambrook *et al.*, 1989). The choice of screening strategy is dependent upon prior knowledge of the gene or its encoded protein:

- **A partial DNA sequence [e.g. a PCR fragment (section 1.2.2) or EST (section 1.2.7)] is known**. The sequence fragment is labelled (commonly using a ^{32}P-labelled nucleotide such as ^{32}P-dCTP), and the library is screened at high **stringency** by hybridisation.
- **A similar DNA sequence (e.g. cDNA of a different species) is known**. The cDNA is labelled as above and the library is screened at low **stringency** by hybridisation.
- **A partial protein sequence (e.g. microsequenced fragment) is known.** Either: (A) **degenerate** oligonucleotides that code for the known amino acid sequence are sequenced, labelled as above, and the library is screened at **low stringency** by hybridisation, or (B) a peptide-specific antibody is raised, and used to screen an **expression library**, in which a

specialised eukaryotic protein-expressing vector is used.

- **A partial protein sequence can be guessed (e.g. by comparison with functionally similar proteins that exhibit conserved domains)**. [see (A) above.]
- **The protein exhibits a specific biochemical/biological activity**. A cDNA expression library is expressed in cells, and a specific biochemical (e.g. ligand binding) or functional (e.g. enzyme activity) assay is used to select clones.
- **The protein can be activated selectively to elicit a second messenger response**. A cDNA expression library is expressed in cells, which are treated with the activator molecule, and second messenger responses are screened (e.g. receptor clones may be activated by specific ligands).

1.2.2 PCR cloning of cDNAs

A revolution in cDNA (and later genomic; see section 1.3.2) cloning that started with the invention of the **p**olymerase **c**hain **r**eaction **(PCR)**, has been coined by Sydney Brenner as 'cloning from the journal *Nature* rather than nature'. PCR is a simple technique for amplifying specific segments of DNA sequence (Box.1.3 and Box.1.4). Coupled with **r**everse **t**ranscription (**RT**) of mRNA the technique of RT–PCR (see Box 6.4) can now be used to obtain novel cDNA sequences. In the absence of a handy source of the appropriate template mRNA, Quick-Clone™ cDNA pools for PCR amplifications are available from commercial sources: **http://www.clontech.com**

BOX 1.3: POLYMERASE CHAIN REACTION (PCR)

First described by Nobel laureate Kary Mullis and colleagues (Saiki *et al.*, 1988), PCR has revolutionised molecular biology, and in particular has enabled progress in the field of molecular neuroscience. The power of the technique lies in the *amplification* of *specific* DNA sequences: starting with either a complex mixture of sequences or alternatively a minute amount of a single sequence, PCR permits the acquisition of a sufficient amount of the desired sequence for cloning, sequencing, quantitation or other experimental purpose.

PCR is dependent upon some prior knowledge of the sequence because specific targets are defined by two flanking **primers** (15–25 bp synthetic oligonucleotides). The usefulness of PCR has been enhanced considerably by clever design of primers. Different categories of primers include:

- Standard primers: where the target sequence is known, primers which complement the 5′ and 3′ ends are easily selected – some design is necessary however (see Box 1.4)
- Conserved sequence primers: PCR may be primed from highly conserved nucleotide sequences such as the poly (A) tail.
- Degenerate primers: PCR may be primed from partially conserved, and predictable, nucleotide sequences that code for conserved protein *domains*. In

Specific if not known arbitrary primers.

BOX 1.3: Contd

this case, the primers are often made **degenerate** in order to accommodate the degeneracy of the genetic code (Box 1.4). This strategy has been used extensively in molecular neuroscience to obtain the sequence of unknown members of multigene families, for example, the G-protein coupled receptor family (see section 1.2.4).

Although a multitude of more sophisticated protocols exists, PCR is basically a simple three-step procedure:

1. Heat DNA template to a **denaturing** temperature (e.g. 94°C for 1 minute) causing DNA strands of the template to separate.

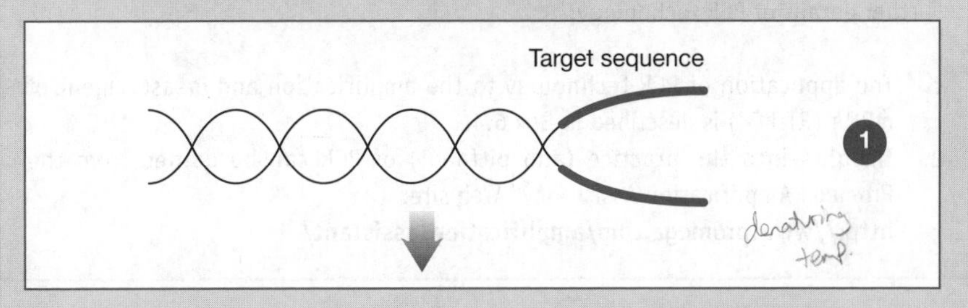

2. Cool to an **annealing** temperature (e.g. 55°C for 1 minute) which allows primers to anneal to 5′ and 3′ ends of the specific target sequence.

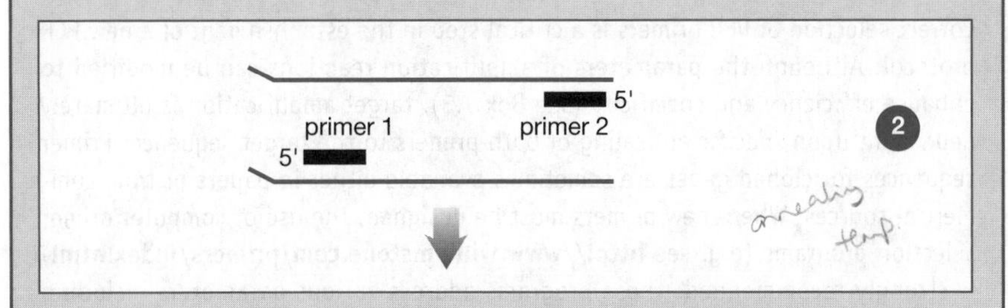

3. Heat to an **extension** temperature (e.g. 72°C for 1 minute) which permits extension of the primers in the presence of the enzyme *Taq* DNA polymerase.

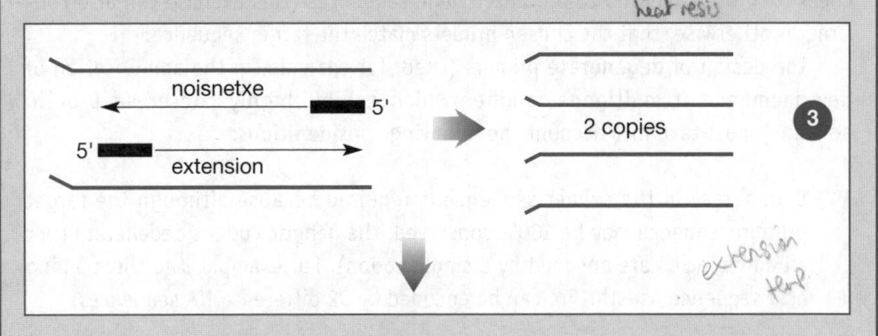

BOX 1.3: Contd

This three-step procedure is repeated (e.g. 30 times) using an automated heating block ('PCR machine') during which time the original template is amplified a million-fold or more.

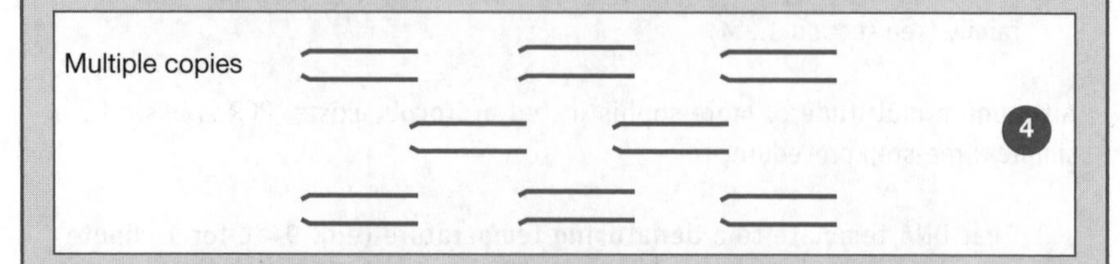

Further details of PCR technology:

1. The application of PCR technology to the amplification and measurement of mRNA (RT-PCR) is described in Box 6.4.
2. Insights into the practice (and pitfalls!) of PCR can be gained from the Promega Amplification Assistant SM Web site: **http://www.promega.com/amplification/assistant/**

BOX 1.4: PCR PRIMER DESIGN

Correct selection of PCR primers is a critical step in the establishment of a new PCR protocol. Although the parameters of amplification reactions can be modified to enhance efficiency and specificity (see Box 1.3), target amplification is ultimately dependent upon specific annealing of both primers to the target sequence. Primer sequences for cloned genes are sometimes available either in papers or from commercial sources. Where new primers must be designed, the use of computer primer selection programs (e.g. see **http://www.williamstone.com/primers/index.html**) is strongly recommended. These programs address various parameters including the **melting temperature** (T_m) of the oligonucleotides, the (related) GC content, and the possibility that the two primers may form stable hydrids with each other. Primer pair sequences should also be run through DNA databases (see Box 1.5) to check for close similarity to unwanted gene sequences (this exercise can also confirm, or otherwise, that the chosen primers match the target sequence!).

The design of **degenerate** primers (used, for example, in the amplification of new members of multigene families which exhibit highly conserved protein domains) must take into account the following considerations:

- Degeneracy in the primer sequence is required because although the target protein sequence may be 100% conserved, the genetic code is degenerate (not all amino acids are encoded by a single codon). For example, the short amino acid sequence Asn-Thr-Pro can be encoded by 32 different DNA sequences:

BOX 1.4: Contd

Asn	*Thr*	*Pro*
AACorAAT	ACAorACCorACGorACT	CCAorCCCorCCGorCCT

In order to ensure that the *single* correct sequence is targeted, it is therefore necessary to use a mixture of primers which encode all of the 32 possibilities – fortunately the introduction of this degeneracy into synthetic oligonucleotide mixtures is a routine procedure.

- It should be apparent that the three amino acid sequence shown above does not represent a sufficiently unique sequence for specific amplification of a PCR product – in practice it has been found that primers should match a DNA sequence encoding at least five amino acids.
- Up to 1024-fold degeneracy in primers is acceptable. The extent of degeneracy is minimised by:
  (i) The 5′ half of primer does not have to be degenerate.
  (ii) The 3′ terminal base of the primers must not match a degenerate position, i.e. the priming site is selected such that the terminal base matches, for example, the second codon position for an amino acid encoded by only 2, 3 or 4 codons.
  (iii) Sequence alignments (for example across different species; see Box 1.5) can be used to demonstrate conservation of particular nucleotides at degenerate codon positions.

Although RT–PCR is widely used in cDNA cloning, it must often be used in conjunction with more time-comsuming techniques such as RACE (see Box 6.5) in order to obtain **'full-length'** sequences that contain the entire coding region of the gene. Commercially available RACE systems can improve the efficiency of this approach (see **http://www.clontech.com**). Alternatively, PCR-derived gene fragments are often used to screen conventional cDNA libraries (see section 1.2.1) for clones containing additional sequence.

1.2.3 Sequence analysis of cDNA clones

The sequence of novel cDNA clones may be obtained using established procedures (see Sambrook *et al.*, 1989), or a variety of newer, automated, sequencing techniques. Subsequent sequence analysis of DNA clones (Box 1.5) is required to determine if:

- the sequence is (a) novel or (b) similar to known genes.
- an **open reading frame (ORF)** is present – the nucleotide sequence codes for a continuous stretch of amino acids, not interrupted by multiple **STOP (translation termination)** codons.

- a **START (translation initiation)** codon (ATG) is present in a functional sequence context (**Kozak consensus sequence**; Kozak, 1996).
- beyond the ORF (in the 3'-terminal region of the cDNA), a **poly (A) addition site** (AAUAAA) signals the terminus of the mRNA which is observed as the poly (A) tail about 30 bases downstream.

BOX 1.5: DNA SEQUENCE ANALYSIS

The explosive growth in DNA sequence data has fortunately been accompanied by the growth of Bioinformatics (see section 2.5), a new scientific discipline which has involved the development of WWW resources to deal with the problem of analysis. General bioinformatics Web sites are described in Box 2.3.

One convenient starting point for DNA sequence analysis is the Baylor College of Medicine (BCM) Search Launcher **(http://gc.bcm.tmc.edu:8088/search-launcher.html)** which has been designed to perform a variety of molecular biology-related searches/analyses from a single point-of-entry:

Data entry

1. Select the required launch page – for cDNA analysis, select *Nucleic acid sequence searches*.
2. Cut sequence data from personal files, and paste onto the submission form. (Alternatively, if the sequence you wish to analyse is already within the database, it may be obtained from the ENTREZ browser **(http://www.ncbi.nlm.nih.gov/Entrez/)** using either a gene name (e.g. enkephalin) or an **accession number** (unique sequence identifier; e.g. M87151)
3. Select from the different search options using a series of buttons. (The BCM launcher contains a variety of sophisticated search tools. An initial analysis of sequence similarity may be performed using BLASTN-nr).
4. Click on *perform search*. (Results are returned within a matter of seconds!)

Results

The output provides a graphical picture of the region(s) of homology within the query sequence, in addition to sequence alignments, i.e.:

```
Query: 5  gcgatgatgctgatctc 21
          || ||||||||| ||||
Sbjct: 14 gccatgatgctgctctc 30
```

A direct link to details of the homologous gene, including the full sequence, may be obtained by clicking on the accesssion number.

BOX 1.5: Contd

Interpretation

Because of the (default) search parameters, and the enormous amount of sequence in the database, you will always get results. The output of a new search will fall into three broad categories:

A. >90% identity
- identical gene (with sequencing errors/mutations, if not 100% identical)
- homologous gene from a related species

B. >70% identity over a significant proportion of new sequence
- homologous gene from an unrelated species
- related gene (i.e. member of same gene family)

C. >60% identity over a significant proportion of new sequence, or >85% identity over a short stretch of bases (e.g. 25)
- these similarities are quite possibly random. The new sequence is not similar to other DNA sequences in the database. Translation of the nucleotide sequence may however reveal significant homology at the protein level.

Further analysis

BEAUTY is an enhanced BLAST-based search tool available at BCM that includes translation of the nucleotide sequence, in addition to post-processing in which annotated protein domain information is provided.

Nucleotide sequence may also be translated at many other Web sites, for example the Swissprot site **(http://expasy.hcuge.ch/www/tools.html)** includes a dedicated translation function that will rapidly provide translation in all six reading frames. Other protein sequence analysis programs at this site include **PSORT** which provides predictions of cellular sorting and localisation signals.

Introductory practical

A web practical (BioActivity) designed to introduce beginners to sequence analysis may be found at:
http://www.biochem.ucl.ac.uk/bsm/dbbrowser/

1.2.4 Applications of cDNA clones

Beyond the mere aquisition of a novel cDNA (which can be a prized and possibly patentable possession), there are multiple applications of the sequence:

- The encoded amino acid sequence can be predicted, and then used to make further functional predictions of protein domains (Box 1.2).

Certain families of genes have individual databases for sequence comparison and analysis (e.g. the G-protein-coupled receptor database: **http://www.gcrdb.uthscsa.edu/** which includes a variety of features including receptor mutations linked to diseases in humans).

- The cDNA sequence can be transcribed and translated *in vitro* (e.g. coupled TNT™ reactions; Promega; **http://www.promega.com**), confirming protein size predictions, and providing protein for functional assays.
- Predicted peptides can be synthesised and used to raise antibodies, which in turn may be used to determine sub-cellular localisation of the protein, and also test for function.
- The cDNA can be **over-expressed** in bacteria (or other systems) to produce protein for further research (e.g. determination of crystal structure) or therapeutics.
- The cDNA can be used as a hybridisation probe for the evaluation of gene expression. The cDNA may either be used directly (e.g. Northern analysis of mRNA levels see Box. 6.2), or transcribed to produce an antisense RNA probe (e.g. *in situ* hybridisation histochemistry; see Box 6.3.).
- The cDNA, or derived oligonucleotides, can be used as hybridisation probes to screen for homologous sequences (**homology screening**) in cDNA libraries (e.g. searching for mutated sequences in human disease state cDNA library).
- The cDNA sequence can serve as the starting point for *in vitro* **forced evolution** studies designed to enhance/modify protein function (e.g. Crameri *et al.*, 1995).

1.2.5 cDNA clones – families and orphans

The discovery of particular cDNAs has facilitated the cloning of multiple related sequences through either homology screening of cDNA libraries (section 1.2.4), or RT–PCR (section 1.2.2). In some cases, such as the nicotinic receptor (see Green *et al.*, 1998), the cloning of a **family** of related genes was unexpected and demanding of many more years of neurophysiological analysis before the individual function of different family members can be elucidated. In other cases, the cognate ligand for a cloned receptor is unknown – such receptors are termed **orphans** (Guest Box p. 13; examples are found within the G-protein-coupled receptor super-family; see section 1.2.4), and their discovery has prompted the search for novel, or not so novel (e.g. cannabinoids; Di Marzo *et al.*, 1994) endogenous ligands. In these ways, molecular cloning is indirectly contributing to our knowledge of functional networks in the brain.

GUEST BOX BY J. P. H. BURBACH

Orphan receptors in the brain

Gene cloning approaches have led to the identification of a number of proteins with homology to proteins of known function. A significant group of these concern 'orphan receptors', proteins that look like receptors, but that lack a known ligand. They are found for all classes of receptor families, but most significantly for nuclear receptors. In fact, the nuclear receptor family now consists of over 70 different genes of which 15 represent true receptors with known ligands. The 50 or more orphan receptor genes are found throughout the animal kingdom with a remarkable degree of conservation.

Nuclear receptors are transcription factors that can be activated by ligands. The classical steroid hormone receptors, e.g. corticosteroid, oestrogen and progesterone receptors, belong to this receptor class, but also receptors for other small, generally lipophilic molecules like retinoids, vitamin D and thyroid hormones act through nuclear receptors. They share a common architecture built of a DNA-binding domain, a transactivating domain and a ligand-binding domain.

There are two theories on the significance of these receptors. The first says that true receptors have evolved from this transcription factor family having acquired the ability to bind small ligands and being modulated by them. The other hypothesises that all orphan receptors have ligands. Based on the latter, a large-scale hunt for the ligands is going on in academic as well as industrial research, driven by the hope to obtain novel compounds for pharmaceutical application. Of several orphans, ligands have indeed been obtained. Many nuclear orphan receptors have a distinct expression pattern in the brain, although they are also expressed in tissues and organs outside the nervous system. A general feature of their expression is the wide and abundant expression during brain development, and a more restricted and lower expression in the mature brain. There is general consensus on the importance of orphan receptors in biological systems like the nervous system. Direct proof has come from *in vivo* studies in which an orphan receptor gene has been inactivated. The null mutation of one orphan receptor, Nurr1, resulted in the failure of progenitor neurones to develop normally into dopamine-producing neurones of the substantia nigra and ventral tegmental area. Aberrant morphology of a cranial ganglion was observed upon genomic mutation of another, COUP-TFI. Notably, in both cases defects were restricted to only a subpopulation of neurones that expressed the orphan receptor. This has been taken as indication that orphan receptors act in tight association with cell-specific events.

The principles of action of orphan receptors are relatively well understood. They regulate the expression of target genes by direct binding to DNA elements. Like most true nuclear receptors, they bind elements composed of a consensus core motif (AGGTCA), but the preference in spacing and orientation of multiple core motifs vary among receptors. This provide some degree of selectivity, but a considerable overlap in target gene recognition exists. Moreover, several orphan

GUEST BOX Contd

receptors can heterodimerise with other orphan or true receptors. Taken together, a complex network of interacting receptors is constituted, thus providing a wide potential to modulate gene expression.

While the importance of ligands in the control of orphan receptor activity is still under debate, other ways for their modulation have been indicated. Several orphan receptors are activated like immediate-early genes by depolarisation or membrane receptor activation. The latter can also result in an altered phosphorylation state of a nuclear receptor, thereby changing its activity. Others claim that orphan receptors have constitutive activity like most other transcription factors. Presently, there can be no common rule applied to all orphan receptors; within an array of options, each will have its own specialities, sometimes depending on the cellular context.

It is clear that current questions need to be answered and will be addressed by loss- and gain-of function studies in genetically modified mice, and likely more simple invertebrate models like *Caenorhabditis élegans* and *Drosophila*. It is to be expected that, in addition to a role during neuronal development, functions in the mature brain at the level of cell physiology will be detailed, in particular by conditional approaches of gene modification. The outlook is that orphan receptors, as participants in molecular cascades of both gene regulation and signal transduction, are in a key position to tune neuronal responses and properties to external conditions.

Lopes da Silva, S. and Burbach J.P.H. (1995) The nuclear hormone-receptor family in the brain: classics and orphans. *Trends in Neurosciences* **18**, 542-548.

J. Peter H. Burbach studied chemistry initially and obtained a PhD in the field of neuropeptide biochemistry at Utrecht University, The Netherlands. Working since then in the Rudolf Magnus Institute for Neurosciences, Utrecht University, his main interest has been the molecular regulation of neuropeptide systems, in particular oxytocin and vasopressin neurones. Part of this research concerns the transcription factors involved in developmental and physiological regulation of these neurones.

1.2.6 Alternative cDNA sequences

One of the surprises of cDNA cloning has been the discovery of alternative cDNA sequences which derive from alternative mRNA forms. Alternative mRNAs may be generated by:

- Alternative transcription start-site.
- Alternative poly (A) addition site – can result in the loss/addition of extensive 3'-terminal untranslated sequence (UTR) and resultant change in mRNA stability.
- **Alternative splicing** – particular exons are removed during splicing, e.g. 'flip' and 'flop' isoforms of glutamate receptor subunits (Sommer *et al.*, 1990). An extreme example of this mode of regulation are the brain-

specific neurexin genes which exhibit more than 1000 isoforms generated by alternative splicing (Ullrich *et al.*, 1995).

- **RNA editing** – the nucleotide sequences of transcripts are subtly modified. Comparison of genomic and cDNA sequences has revealed the presence of nucleotide substitutions, for example in the GluR2 glutamate receptor sub-unit (see Green *et al.*, 1998) which alter coding potential, and functional activity of the protein (also see Further Reading and the Guest Box on p. 35).

1.2.7 Partial cDNA sequences – ESTs

Expressed sequence tags (ESTs) are partial cDNA sequences, randomly generated by RT–PCR (section 1.2.2) using non-specific primers. Large numbers of ESTs may be rapidly cloned in this manner, providing a source of both novel gene sequences, and genetic markers (section 2.3.1). The use of specific tissues as a source of template mRNA (e.g. brain; Adams *et al.*, 1993) has enabled comparisons of tissue-specific EST expression patterns.

There are now millions of (overlapping) ESTs in the database (see section 2.3.1), of which the majority are human sequences, although other species including mouse and rat are well represented. It is estimated that ESTs now cover >50% of all human genes. Because the full sequence of many of these genes is unknown, ESTs are a valuable aid to gene discovery (see, for example Box 1.6).

BOX 1.6: *IN SILICO* CLONING

More than one million ESTs, which are randomly cloned cDNA fragments (see section 2.3.1), are now stored in the public EST sequence database (dbEST; 2.4.1). Although often just a few hundred base pairs in length, the ESTs are a valuable aid to gene discovery because they are easily obtained (e.g. from Research Genetics: **http://www.resgen.com**) and used, for example, as probes for DNA library screening (section 1.2.1). In this way 'full-length' cDNAs may be obtained and used for further research.

The amassing of (overlapping) EST sequences has also enabled *'in silico'* cloning. This refers to the use of specialised computer search tools which can be used to assemble related ESTs into contiguous gene sequences. An example is ESTblast which is available on the Human Genome Mapping Project (HGMP) Resource Centre **(http://www.hgmp.mrc.ac.uk/ESTblast)**.

Data entry

1. Either: (a) Cut sequence data from personal files, and paste onto the submission form.
   Or: (b) Use the Fetch option to obtain the query sequence (accession number required).

BOX 1.6: Contd

2. Select the program to use.
   (BlastN will provide a nucleotide sequence search against the nucleotide database; TBlastN searches protein sequence against the nucleotide database.)
   (Other options can be left on default.)
3. Click on *GO*!
   (Results are returned within seconds!)

Results

The first search output identifies similar ESTs, and also provides a sequence alignment (see Box 1.2).

(A direct link to details of the homologous gene, including the full sequence, may be obtained by clicking on the accesssion number.)

4. Click on *Create multiple alignment*.
   The next output allows selection of similar ESTs for the subsequent multiple alignment .
5. After selecting the ESTs, click on *Alignment*.
   The next output provides a graphical view of the overlapping EST sequences:

```
GRAPHICAL VIEW OF ESTblast

[ ] H75648                         =============><————————— H75649 [ ]
[ ] AA54678   ===============><—————                       AA54670 [ ]
[ ] Query              =========>                          No data [ ]
[ ] No data                  <=============                 D54677 [ ]
```

The output has been simplified to illustrate the essential features of these searches. ESTs are represented by arrows. (==) Both sequence and position are known. (—)Known sequence, unknown position. Note that the program obtains the ESTs from the other end of the clone (e.g. H75648 and H75649) if available, or if not (No data) is indicated. Note that the original query sequence has been significantly extended in both directions. Using the flanking boxes, sequences may be selected for subsequent consensus construction.

6. Click on *CREATE*!
   This provides a consensus sequence of the overlapping ESTs. It is then possible to 'Re-Blast' the '*in silico*-cloned' consensus sequence and thereby obtain further related sequences.

Interpretation

Like other virtual reality situations, *in silico* cloning cannot yet replace the real thing:

- Because of the priming strategy , many ESTs are representative of the 3' termini of mRNAs only, and therefore it may not be possible to 'clone' 5' terminal sequence (back to the lab!).

BOX 1.6: Contd

- Sequences that display homology over short overlaps may not be representative of naturally occuring mRNAs, and therefore false consensus sequences may be constructed.

The National Center for Biotechnology Information (NCBI) routinely prepares clusters of ESTs that belong to the same transcript. These may be sourced at: **http://www.ncbi.nlm.nih.gov/UniGene/index.html.**

1.3 Genomic DNA cloning

Unlike cDNA (section 1.2), which is representative of only the expressed portion of genes, genomic DNA is fully representative of chromosomal DNA sequence – it contains both exons and introns in addition to regulatory sequence.

1.3.1 Genomic DNA libraries

Genomic DNA has traditionally been cloned by cutting chromosomal DNA into random fragments [e.g. 15–20 kilobases (kb) long] using rare-cutting **restriction enzymes**; genomic libraries are then constructed (Sambrook *et al.*, 1989) and screened, often using cDNAs as a probe. Individual laboratories rarely produce their own genomic libraries; it is common practice to either obtain an aliquot of a library from a friendly laboratory, or alternatively purchase a library (e.g. Stratagene; section 1.2.1). There are additional sources of genomic DNA, and now genomic sequence may be obtained as:

- bacteriophage clone (up to 20 kb; e.g. library from Stratagene)
- cosmid clone (up to 45 kb; e.g. library from Stratagene)
- bacterial artificial chromosome (BAC; up to 150 kb; e.g. from Research Genetics Inc.: **http://www.resgen.com**)
- yeast artificial chromosomes (YAC; up to 1 Mb; e.g. from Research Genetics Inc.)

The choice of these alternatives is dictated by research requirements. Clearly, larger clones such as BACs could be anticipated to contain entire 'genes' ('**transcription units**') which would encompass all regulatory sequences in addition to coding and non-coding exons, and introns. However, such large clones may well contain sequence from more than one gene. The use of BACs in novel gene identification is discussed later (see section 8.3.2.). The provision and screening of genomic clones is a rapidly developing science/commercial enterprise (see News at Research Genetics Inc.).

1.3.2 PCR cloning of genomic sequence

PCR gene amplification techniques (Box 1.3) have recently been improved to enable the amplification of large (>20 kb) stretches of DNA from either genomic clones or tissue-extracted genomic DNA. This approach can provide a short-cut to both genomic sequence analysis and genomic cloning, but is clearly limited by the availability of sequence (e.g. cDNA) for primer design. Many molecular biology companies have invested in this technology (e.g. Expand™ PCR sytems from Boehringer Mannheim; **http://biochem.boehringer-mannheim.com.**). Kits containing premade libraries for the cloning of unknown genomic sequences are also available (GenomeWalker™; **http://www.clontech.com**).

1.3.3 Sequence analysis of genomic clones

Genomic sequence may be checked for similarity using sequence search tools (Box 1.5) if the purpose of the cloning is to identify novel genes; a variety of other forms of analysis are also used for gene identification in large genomic clones (see section 8.3.2; King *et al.*, 1997).

Specialised search tools have been developed for the analysis of gene promoter regions. For example, TESS (transcription element search software; **http://agave.humgen.upenn.edu/utess/tess31**) answers the question: 'What potential transcription factor binding sites are located in my sequence and where are they?' The results of this type of search are, however, no substitute for biological assays of transcription factor binding (see Figure 7.4).

1.3.4 Applications of genomic clones

- Definition of gene structure, i.e. exon/intron boundaries.
- Sequence analysis of promoter region.
- Analysis of promoter function *in vitro* (e.g. by using **promoter/reporter gene** constructs in transfected cell lines).
- Analysis of promoter function *in vivo* (e.g. by using transgenic mice; see Box 4.3).
- Construction of gene targeting constructs (see section 4.3.3).

1.4 Summary

- Complementary (c)DNA copies of mRNAs are cloned into vectors for sequencing, expression, and other studies.
- cDNA libraries are screened using a variety of different approaches.
- PCR-based cDNA cloning is increasingly used but often does not fully substitute for conventional library screening.

- Sophisticated, computer-based sequence analysis packages are now available on the World Wide Web.
- cDNA cloning has revealed the presence of multigene families, and also orphan molecules which have no recognised functional role.
- Alternative mRNA sequences are derived through a variety of mechanisms including alternative RNA splicing and also RNA editing.
- The cloning of a multitude of partial cDNA sequences (ESTs) has formed an important resource for gene discovery – *in silico* cloning may be used to derive full-length cDNA sequences.
- Genomic DNA clones are required for a variety of applications (where cDNA sequence is insufficient) including promoter analysis, and transgenic animal production.

Further reading

HALL, S.S. (1987) *Invisible Frontiers: The race to synthesize a human gene.* Tempus Books, Washington, USA.

A journalistic account of the early days of gene cloning which succeeds in conveying the 'ups and downs ' (in this case, major) of laboratory research.

SIMPSON, L. and EMESON, R.B. (1996) RNA editing. *Annual Review of Neuroscience*, **19**, 27–52.

RNA editing is increasingly recognised in neuronal gene transcripts and therefore deserves attention in this annual review of progess in neuroscience.

MONYER, H. and LAMBOLEZ, B. (1995) Molecular biology and physiology at the single-cell level. *Current Opinion in Neurobiology* **5**, 382–387.

PCR technology has been successfully applied at the level of single neurones using 'Patch-PCR' and other novel approaches.

CHAPTER TWO

Why not sequence everything?

Key topics

- Genome projects
- The mammalian genome is full of junk
- Finding genes in the junk
- Which genes are expressed in the brain?
- Bioinformatics
- Functional genomics

2.1 Introduction

The cloning of the genes encoding proteins that are responsible for known neuronal activities can yield important information about the structure, function and regulation of these moieties (see Chapter 1). This approach has also yielded tantalising evidence of the complexity of the **genome**, and the diversity of the gene families that it encodes. However, this is but a tiny part of the story. Only a small fraction of the genes expressed in the brain have thus far been identified, characterised and studied.

Mammals are thought to have about 70 000 genes, a subset of which, maybe 50 000, are expressed in the brain. The identification of these genes is a necessary prelude to any attempt to construct a model of brain function based on the integrated activity of gene networks. Such a global approach demands that we identify all of the genes that are expressed in the mammalian brain, that we determine precisely when and where in the brain these genes are expressed and, finally, that we determine their functions.

2.2 Genome projects

Large-scale projects are being undertaken to map and sequence entire genomes from a number of different evolutionary lineages (Table 2.1). The complete sequences of the *Escherichia coli* (see the *E. coli* Genome Project – **http://www.genetics.wisc.edu/index.html**) and yeast (*Saccharomyces cerevisae*; see the *Saccharomyces* Genome Database – **http://genome-www.stanford.edu/Saccharomyces/**) genomes are already available. Soon, the first sequence of a metazoan organism with a nervous system, that of the nematode, *Caenorhabditis elegans*, will be completed. However, the most ambitious projects involve the mammalian genomes, which span 4000 megabase pairs (Mb).

Table 2.1 Genome project goals.

Goal	*Resolution*
Complete a detailed human genetic map	2Mb
Complete a physical map	0.1 Mb
Acquire the genome as clones	5 kb
Determine the complete sequence	1 b

The aim of the various Genome Projects is to make a series of maps of each chromosome at increasingly finer resolution (Figure 2.1; Table 2.2). Gene maps are made by a combination of **genetic** and **physical** methodologies.

2.2.1 Genetic linkage maps

A **genetic** linkage map describes the order of genes on a chromosome. Any inherited physical or molecular characteristic that differs between individuals is a potential genetic marker. DNA sequence differences are particularly easy to characterise. Such markers must be **polymorphic** to be useful in mapping – alternative forms must be present in different individuals that can be followed in family studies. Most polymorphisms have no detectable effect on the function or appearance of an organism, but they can be readily monitored at the DNA level. Examples of these types of markers include:

- **restriction fragment length polymorphisms (RFLPs)**, which correspond to variations in DNA sequences that give rise to restriction enzyme cleavage sites that are present in some individuals, but not others.
- **simple sequence length polymorphisms (SSLPs)**, which correspond to variations in the lengths of simple sequence repeats (SSRs). SSRs are repetitive sequences that vary in the number of simple repeated units and, therefore, in length.

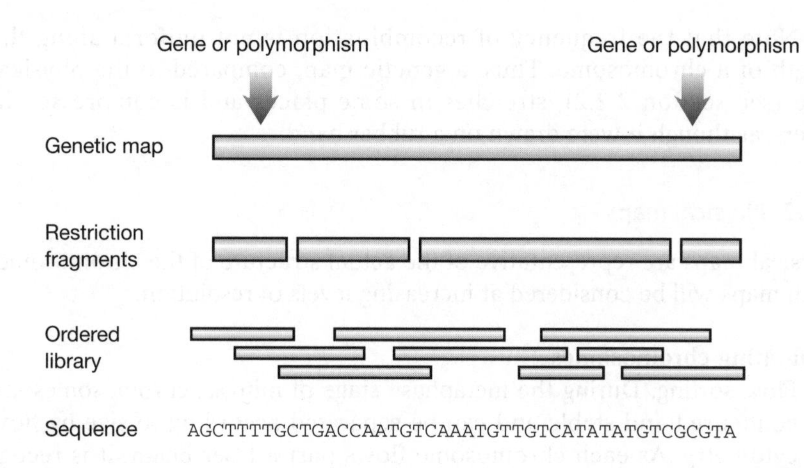

Figure 2.1 At the coarsest resolution, the genetic map measures recombination frequency between linked markers (genes or polymorphisms). At the next resolution level, restriction fragments of 1 to 2 Mb can be separated and mapped. Ordered libraries of cosmids and YACs have insert sizes from 40 to 400 kb. The nucleotide sequence is the ultimate physical map.

A genetic linkage map is constructed by measuring the frequency with which two markers are inherited together. Two markers on the same chromosome can segregate as a consequence of meiotic recombination. The closer the markers are to each other on the chromosome – that is, the closer the linkage – the less likely is a recombination event that will separate them. **Recombination frequency** thus provides an estimate of the distance between two markers. Genetic map distances are measured in centimorgans (cM), named after the American geneticist Thomas Hunt Morgan. Two markers are 1 cM apart if they separate by recombination 1% of the time. A genetic distance of 1 cM is roughly equivalent to a physical distance of 1 Mb.

Table 2.2 Gene number and genome size in organisms in different evolutionary lineages.

		Genes	*Genome size in Mb*
Prokaryota	*Escherichia coli*	4100	4.7
Fungi	*Saccharomyces cerevisiae*	6300	13.5
Arthropoda	*Drosophila melanogaster*	12 000	165
Nematoda	*Caenorhabditis elegans*	14 000	100
Chordata	*Fugu rubripes*	70 000	400
	Mus musculus	70 000	3300
	Homo sapiens	70 000	4000

Note that the frequency of recombination is not uniform along the length of a chromosome. Thus, a genetic map, compared to the physical map (see section 2.2.2), stretches in some places and is compressed in others, as though it were drawn on a rubber band.

2.2.2 Physical maps

Physical maps are representative of the actual structure of the chromosome. Such maps will be considered at increasing levels of resolution.

Separating chromosomes

- Flow sorting. During the metaphase stage of mitosis, chromosomes are condensed and stable and can be separated according to size by flow cytometry. As each chromosome flows past a laser beam, it is recognised by analysing the amount of DNA present, and is sorted into a specific collection tube.
- Somatic cell hybridisation. Fused human and rodent tumour cells preferentially lose human chromosomes until only one or a few remain. Hybrid cell clones containing specific human chromosomes can then be propagated.

Cytogenetic maps

Cytogenetic maps assign the positions of specific DNA markers to particular bands on chromosomes identified by staining. The DNA marker is tagged with a fluorescent or radioactive label, and is hybridised to metaphase chromosomal spreads. The location of the labelled probe can be detected after it binds to its complementary DNA strand in an intact chromosome.

Macrorestriction maps

A macrorestriction map describes the order and distance of restriction enzyme cleavage sites on a chromosome, allowing DNA markers to be located at a resolution down to 100 kb. The development of **pulsed-field gel electrophoresis** (PFGE) has enabled the mapping and cloning of large DNA molecules; while conventional gel electrophoretic methods can separate fragments up to 40 kb, PFGE can separate molecules up to 10 Mb because periodic changes (pulses) in the direction of the electric field enhance the resolution of DNA fragments.

Contig maps

A **Contig** map depicts the order of overlapping cloned DNA fragments that span the length of a chromosomal segment. Chromosomes are cut into small pieces, which are cloned into cosmid, **BAC** or **YAC** vectors (see section 1.3.1), and ordered. The ordered fragments form contiguous DNA blocks (contigs). Contigs can be verified by cytogenetic mapping.

Sequencing
The highest resolution physical map is the complete elucidation of the DNA sequence of each chromosome of the genome. Sequencing requires, firstly, the subcloning of small fragments of large genomic clones into special sequencing vectors. The next step is to represent the subcloned fragments as a set of nested fragments differing in length by one nucleotide, so that the specific base at the end of each successive fragment is detectable after separation by gel electrophoresis.

2.3 Junk DNA

Over 90% of the mammalian genome is made up of 'junk', repetitive elements that have no known function either as sequences that code for proteins or regulate gene expression. The sequencing of mammalian genomic DNA is thus an inefficient way of obtaining information - effectively, for every 100 base pairs of genomic DNA that are sequenced, only 10 base pairs correspond to information coding for proteins. Two approaches are being used to get around the junk problem: firstly, to sequence only that part of the genome that is expressed; and secondly, to study model organisms that are devoid of junk.

2.3.1 Expressed sequences

The expressed sequences of a particular cell or tissue are represented by the **messenger RNA** (mRNA) pool, which can be isolated and used as a template to synthesise complementary or copy DNA (cDNA). This can be cloned and sequenced (see Chapter 1). Partial sequences of cDNAs are called **expressed sequence tags (ESTs;** see section 1.2.7 and **http://www.ncbi.nlm.nih.gov/dbEST/index.html**). ESTs, as well as identifying expressed sequences, also provide unique markers for genome mapping.

A recent technological advance - **Serial Analysis of Gene Expression** (**SAGE**; see Box 2.1) - enables the rapid and efficient qualitative and quantitative description of the expression pattern of a large number of genes in any given tissue or group of cells.

2.3.2 *Fugu rubripes* – a junk-free vertebrate

The pufferfish, *Fugu rubripes*, has a compact genome of 400 Mb that contains very few repetitive elements, yet this vertebrate has a gene repertoire comparable with that of mammals (Table 2.2). The majority of introns in the pufferfish are small (modal value 80 b) and repetitive sequences account for <10% of the genome. Thus, the pufferfish genome is an attractive model for identifying and characterising genes in the context of overall chromosome structure.

BOX 2.1: SERIAL ANALYSIS OF GENE EXPRESSION (SAGE)

SAGE is a patented, proprietary, sequence-based technology for gene identification a quantitation developed by Bert Vogelstein, Ken Kinzler and Victor Velculescu at the Joh Hopkins University Oncology Center (**http://www.sagenet.org**). SAGE is marketed Genzyme Corporation (**http://www.genzyme.com/prodserv/molecular_oncology/sag welcome.htm**). SAGE is highly efficient, able to detect low abundance genes, very accurat and extremely sensitive. The SAGE technique enables the rapid and efficient qualitative a quantitative description of the expression pattern of a large number of genes in a giv tissue or cells. The method is based on two principles that enable rapid, accurate analysis gene expression patterns:

1. Short oligonucleotide sequences from a defined location within a transcript ('tag') all accurate quantitation. A short (9–10 b) nucleotide sequence is sufficiently specific to di criminate different transcripts from each other. For example, it has been estimated th 9-b sequences can distinguish 262 144 different transcripts. Given that the estimat number of human genes is lower than 100 000, the 9-b sequences can provide informati on expression of all human genes.
2. Concatenation of the short sequence tags allows efficient analysis of transcripts within single clone. Sequencing of the concatenated short sequence tags in the clones follow by sequence comparison with the nucleotide databases identifies expressed sequence The number of times that a sequence of a given gene is found is a measure of the level expression of the transcript.

   Double-stranded cDNA is synthesised from any mRNA using reverse transcriptase and biotinylated oligo(dT) primer, which anneals to mRNA poly (A) tails. The cDNAs are cleav with one of the restriction endonucleases (anchoring enzyme) that recognises a 4 b reco nition site. These restriction endonucleases cleave, on average, every 256 b, therefo most transcripts are cut more than once. The most 3' part of the cleaved biotinylated cD is bound to strepavidin beads. This assures that the site binding to the streptavidin closest to the poly (A) tail. The cDNA is divided in half and ligated by way of the anchori restriction enzyme site to one of two linkers that contain a type IIS restriction site (ta ging enzyme). Type IIS restriction endonucleases cleave at a defined distance of up to 2 from their asymmetric recognition sites. The linkers are designed so that cleavage of t ligation products with the tagging enzyme leads to the release of the linker along with short fragment of cDNA. The two pools of released tags are ligated to each other by t created blunt ends. This is followed by PCR to amplify the tag sequences and provide o entation and punctuation. The amplified product contains the two tags (ditags) which a linked tail-to-tail and are flanked by sites for the anchoring enzyme. This results in 4 punctuation per ditag. Then, the ditags are released from the PCR products by cleava with an anchoring enzyme. The released ditags are concatenated by ligation, cloned a sequenced. Specially designed software is used to perform tag identification and quant tation, and to create a relational database of the expression profile.

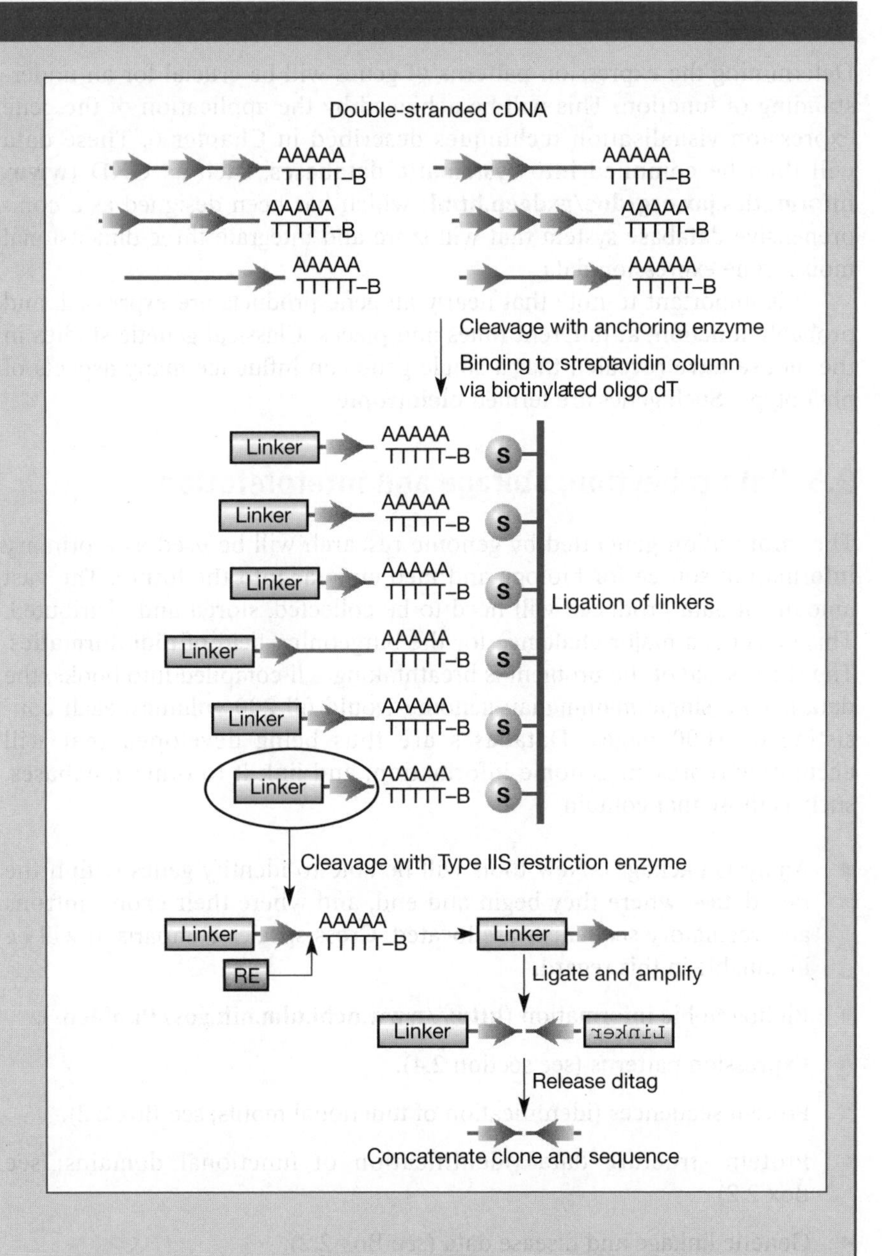
Double-stranded cDNA
AAAAA
TTTTT–B
Cleavage with anchoring enzyme
Binding to streptavidin column
via biotinylated oligo dT
Linker
S
Ligation of linkers
Cleavage with Type IIS restriction enzyme
RE
Ligate and amplify
Release ditag
Concatenate clone and sequence

2.4 Which genes are expressed in the brain?

Determining the expression patterns of genes will be crucial for an understanding of function. This will be achieved by the application of the gene expression visualisation techniques described in Chapter 6. These data will then be compiled into systematic databases, such as GXD (**www.informatics.jax.org/doc/gxdgen.html**) which has been designed as a comprehensive database system that will store and integrate three-dimensional mouse gene expression data.

It is important to note that nearly all gene products are expressed, and probably function, at different times and places. Classical genetic studies in the mouse have revealed that a single gene can influence many aspects of phenotype. Such genes are termed **pleiotropic**.

2.5 Data collection, storage and interpretation

The information generated by genome research will be used as a primary information source for biology and medicine far into the future. The vast amount of data produced will need to be collected, stored and distributed. This presents a major challenge for the burgeoning field of **bioinformatics**. The sheer scale of the problem is breathtaking – if compiled into books, the data from a single mammalian genome would fill 200 volumes, each consisting of 1000 pages. Databases are thus being developed that will accurately represent genome information, and link it to other databases, such as those that contain:

- Analysis packages. New tools will be able to identify genes within the raw data – where they begin and end, and where their exons, introns and regulatory sequences are located. Cross-species comparison will be invaluable in this regard.
- Bibliographic information (**http://www.ncbi.nlm.nih.gov/PubMed/**).
- Expression patterns (see section 2.4).
- Protein sequences (identification of functional motifs; see Box 2.2).
- Protein structure data (identification of functional domains; see Box 2.2).
- Genetic linkage and disease data (see Box 2.2).

It still amazes the present authors, who are old enough to remember punch cards, that all of this information will be instantly available on your desktop computer, via the World Wide Web. Already, the Internet enables access to some amazing bioinformatics sites (see Box 2.2).

BOX 2.2: BIOINFORMATICS ON THE WORLD WIDE WEB

Mapping databases

- The Genome Data Base (GDB), located at Johns Hopkins University (Baltimore, Maryland), provides location, ordering and distance information for human genetic markers, probes and contigs linked to known human genetic disease. GDB is presently working on incorporating physical mapping data.
  http://wwwtest.gdb.org/
  http://www.hgmp.mrc.ac.uk/Courses/gdb/5day.top.html
  http://wwwtest.gdb.org/gdb/contact.html#nodes
- The Online Mendelian Inheritance in Man database is a catalogue of inherited human traits and diseases. OMIM also originated from Johns Hopkins, and was developed for the World Wide Web by the NCBI (National Center for Biotechnology Information). **http://www.ncbi.nlm.nih.gov/Omim/**
- The Mouse Genome Database (MGD) contains information on mouse genetic markers, molecular segments, phenotypes, comparative mapping data, experimental mapping data and graphical displays for genetic, physical and cytogenetic maps.
  http://www.informatics.jax.org/doc/aboutMGI.html
- An interactive mouse genome viewing facility.
  http://www.mpimg-berlin-dahlem.mpg.de/~andy/GN/mitmouse/
- Genetic and physical maps of the mouse.
  http://www.genome.wi.mit.edu/cgi-bin/mouse/index
- Mouse cytogenetics.
  http://ws4.niai.affrc.go.jp/dbsearch2/mmap/mmap.html
- Rat genome mapping.
  http://ratmap.gen.gu.se/
  http://www-genome.wi.mit.edu/ftp/ distribution/ rat_sslp_releases/

Sequence databases

Nucleic acids

Public databases containing the complete nucleotide sequence of the human genome and those of selected model organisms will be one of the most useful products of the Genome Projects. Four major public databases now store nucleotide sequences:

1. GenBank – **http://www.ncbi.nlm.nih.gov/Web/Genbank/index.html**
2. Genome Sequence DataBase (GSDB) – **http://www.ncgr.org/gsdb/**
3. European Molecular Biology Laboratory (EMBL) Nucleotide Sequence Database – **http://www.ebi.ac.uk/ebi_docs/embl_db/ebi/nar/gkb052_gml.html**
4. DNA Data Bank of Japan (DDBJ) – **http://www.ddbj.nig.ac.jp/htmls/Welcome-e.html**

The databases collaborate to share sequences, which are compiled from direct author submissions and journal scans.

BOX 2.2: Contd

Proteins

The major protein sequence databases are:

1. Protein Identification Resource (National Biomedical Research Foundation) – **http://www.psc.edu/general/software/packages/nbrf-pir/nbrf.html**
2. Swissprot – **http://expasy.hcuge.ch/sprot/sprot-top.html**

Sequence similarity searching

The amazing BLAST (Basic Local Alignment Search Tool) sequence similarity search is located at **http://www.ncbi.nlm.nih.gov/BLAST/**

Protein structures

The Molecular Modelling Database (MMDB) contains three-dimensional structures determined by X-ray crystallography and NMR spectroscopy – **http://www.ncbi.nlm.nih.gov/Structure/index.html**

The Protein Data Bank (PDB) is an archive of experimentally determined three-dimensional structures of proteins – **http://www.pdb.bnl.gov/**

2.6 What do all of these genes do?

Information in databases will not, by itself, be sufficient to determine biological function, but it will provide an important foundation for the design of appropriate experiments. The remarkable wealth of information that molecular genetics has provided us with needs to be integrated into an understanding of the functioning of whole tissues, organs and organisms. Without such integration, molecular information is nothing more than a confusing catalogue of sequences and structures. But how do we determine the function of each of the 70 000 genes of the mammalian genome, and particularly of the 50 000 genes that are expressed in the brain? This is the daunting task of the new discipline of **functional genomics**. Functional genomics will rely on the comparative study of genes from different **model organisms.**

2.6.1 Model organisms

The 70 000 genes of the human genome encode components of a much smaller number of core biological processes of known biochemical function, such as amino acid biosynthesis, protein synthesis, protein secretion and cell cycle regulation. These components, and the way that they interact with each other, will be conserved through evolution. Furthermore, the number of core biochemical processes is likely to be similar in all metazoans.

The experimental tools exist in the model organisms, but not in humans, for assembling genes into pathways. Particularly, **transgenic** technologies applied to mouse and rats (Chapter 4) enable specific loss-of-function, over-expression or mis-expression phenotypes to be generated.

2.6.2 Comparative genomics

One approach to the evaluation of gene function and regulation is to identify evolutionarily conserved sequences and structures by interspecies comparisons. Conservation of function also occurs at higher levels. In many cases not only individual protein domains and proteins, but entire multisubunit complexes and biochemical pathways are conserved. In many cases, the way in which these complexes and pathways are utilised in the development and physiology of the organism are also conserved.

2.6.3 Protein structure

The function of a gene is dependent upon the structure of the protein that it encodes, and, in turn, on specific interactions of that protein with substrates, cofactors and other proteins. As yet, it is not possible to predict directly the three-dimensional structure of a protein from primary sequence information. However, it is possible to identify functional domains that are conserved between proteins from the same or different species. Such domains may, in some cases, be recognised in the primary sequence. However, as it is more likely that three-dimensional structures are more conserved than amino acid sequences, analytical capabilities will need to be developed that facilitate grouping protein sequences into motif families. Comparison of genome databases from different species will thus provide a basis for assessing the conservation of functional domains, proteins and multisubunit complexes.

2.6.4 Transgenesis

Transgenesis techniques (see Chapter 4) that enable the transfer of genes between species, allow similarities identified at the levels of sequence and structure to be directly tested at the level of physiological function.

Genome project data and transgenic experimentation data will determine the extent to which genes from different organisms can be swapped around, but still retain function. Studies on mice and rats bearing *Fugu* transgenes have clearly demonstrated that teleost genes can be expressed in a mammalian system under the control of their own regulatory sequences. For example, the *Fugu* gene encoding the neuropeptide isotocin is expressed in rat brain in a cell-specific manner, and responds to physiological stimuli like the equivalent rat gene (Venkatesh *et al.*, 1997). High gene density in the *Fugu* is an additional advantage in transgenic studies. The average gene density in *Fugu* is about one gene per 6 kb. Thus, a *Fugu* genomic cosmid contains about seven complete genes together with their regulatory elements. By using *Fugu* cosmids for generating transgenic rats, clusters of *Fugu* genes can be introduced simultaneously into the rat genome and their expression pattern analysed. Transgenics bearing clusters of genes would be particularly useful in identifying elements that are involved in the locus regulation and the correct expression of genes. In the expanded genome of mammals, such locus control elements are normally dispersed over long genomic regions.

The generation and analysis of transgenic rats bearing *Fugu* genes represents a general approach for identifying and characterising conserved regulatory elements. Although it may be possible to express *Fugu* genes in other teleosts such as zebrafish, such studies will not be useful in localising regulatory elements, as large stretches of physiologically non-significant intergenic sequences might be conserved between these two closely related teleosts. In contrast, as *Fugu* and rat are separated by 400 million years, non-significant sequences would have had ample time to randomise by mutation and translocation. Important conserved sequences, that have maintained functions in gene expression and regulation, can thus be readily identified.

2.7 Perspectives

The new millennium will see the completion of the mapping and sequencing of the human genome, and of the genomes of numerous model organisms. Analytical tools will be developed that enable this information to be related to expression and structural information. However, even in this integrated form, the usefulness of databases will be limited. This is because the components of biological systems function in a non-linear fashion; activities are usually controlled synergistically rather than as simple on/off switches. For example, a small change in the concentrations of transcriptional components can result in a large response. Databases cannot predict such non-linear responses. Rather, complex biological systems – particularly brain function – will only be understood through experimentation *in vivo*. Increasingly sophisticated transgenic manipulations involving the shuttling of genes and regulatory sequences between model organisms, will be crucial in this regard (see Chapter 4).

2.8 Summary

- Large-scale genome projects will identify and sequence every human gene.
- The expression pattern of each gene in the brain will be determined.
- The major challenge for the emerging field of functional genomics is to integrate this catalogue of gene sequences, protein structures and gene expression patterns into an understanding of physiological function.
- Transgenesis will be used to swap genes between model organisms and will allow similarities identified at the levels of sequence and structure to be directly tested.

CHAPTER THREE

Genetic mutations

Key topics

- Types of DNA sequence mutations
- Molecular misreading
- Molecular effects of mutations
- Aberrant RNA processing in neurones
- Mutation analysis in animal models
  - Directed mutagenesis
  - Insertional mutations
  - Random mutagenesis
  - Mutagenesis screens
  - Screening procedures
  - Mouse mutants and the genetics of hearing loss
- Identification of mutant genes
  - Candidate genes
  - Positional cloning

3.1 Introduction

Genes are termed **mutant** where the DNA sequence differs from the normal or **wild-type** sequence. Genetic mutations may be either:

- **Functional**: the function of the gene is changed by:
  (A) a loss-of-function mutation – wild-type gene function is lost through altered mRNA or protein. It is important to note that single base changes can have profound effects on **phenotype** – a single A→T change in the mouse *clock* gene drastically alters circadian behavioural rhythms (see section 8.3).

(B) a gain-of-function mutation – a novel functional activity is generated through altered mRNA or protein.

Both loss-of-function and gain-of-function mutations are recognised causes of neurological diseases (see Box 9.2).

- **Silent**: the DNA sequence change does not alter gene function. For example, a single DNA base change may either not alter the coding potential of the gene (the genetic code is degenerate; Table 3.1), or the change may result in a conservative change in amino acid sequence that has no functional consequences.

There are two reasons for studying mutant genes:

- Gene mutations form the molecular basis of genetic diseases.
- Specific mutations provide insights into the structure and function of genes/RNAs/proteins.

3.2 Molecular classification of mutations

DNA mutations are of two types:

- Length mutations – these involve gain or loss of genetic material. They can take a variety of forms including deletions, duplications and insertions. Trinucleotide repeat expansions are also an important example of length mutations, forming the molecular basis of a group of neurodegenerative diseases that includes Huntington's disease (see section 9.4.1).
- Point mutations – here, the sequence is altered, but there is no gain or loss of genetic material. Different types of point mutations are illustrated in Table 3.1

3.2.1 Molecular misreading

An important new class of mutation termed molecular misreading has recently been identified in which mutant RNAs are transcribed from unaffected DNA (see Guest Box p. 35).

Table 3.1 DNA mutations.

DNA sequence	*Amino acid sequence*	*Mutation*
GTC TTC CGA CGA	Gln-Lys-Ala-Ala	Normal sequence
GTC TT**T** CGA CGA	Gln-Lys-Ala-Ala	Silent point mutation
GTC **C**TC CGA CGA	Gln-Glu-Ala-Ala	Missense point mutation
GTC **A**TC CGA CGA	Gln-STOP	Nonsense point mutation

GUEST BOX BY F.W. VAN LEEUWEN

Different types of mutations in the genome and transcripts of the Brattleboro rat

The Brattleboro rat exhibits a Mendelian inherited diabetes insipidus (di) due to a mutation in the single autosomal gene for the hormone vasopressin. Since both alleles for the gene are transcribed, the trait is recessive. The mutation is a single nucleotide deletion (G) in the second exon of the gene. As a result of this frameshift mutation, a precursor protein with an aberrant C-terminus is formed that is arrested in the membranes of the endoplasmic reticulum and thus unable to enter the secretory pathway. Consequently, a severe diabetes insipidus develops: about 70% of the body weight is excreted daily as hypotonic urine. Remarkably, it was discovered that in the Brattleboro rat a low but increasing number of cells are generated throughout the animal's life that are able to synthesise, transport and release vasopressin into the circulation. However, the amount of vasopressin synthesised is far too low to recover from di. Molecular cloning has made it possible to establish that the cells with such a revertant vasopressin phenotype underwent a novel type of mutation in their transcripts (thus not in the genome!). A dinucleotide deletion (GA) in the mRNA downstream of, and in combination with, the upstream located single base (G) germline mutation results in the restoration of the wild-type reading frame. The resulting vasopressin precursor (a stretch of 13 or 22 amino acids is still mutated) is admitted to the secretory pathway and processed into biologically active vasopressin.

By an unknown mechanism the RNA mutations (GA) arise from unaffected DNA during transcription. This process, designated 'molecular misreading', was also shown in wild-type rat and human vasopressin transcripts. However, here, contrary to Bratleboro rats, proteins within an aberrant C-terminus are formed. Such a dinucleotide deletion in the open reading frame of an mRNA molecule leads to translation of a protein with a wild-type N-terminus, but at the site of the RNA mutation the reading frame is shifted. As a result, the C-terminus is translated in the +1 reading frame, which leads to the synthesis of a so-called '+1 protein'. These dinucleotide deletions (GA) in transcripts occur in or near simple repeats such as GAGAG, which are apparently hot-spots for mutations. The transcript mutation rate was shown to be dependent on the transcriptional activity. The mutations often generate downstream premature termination codons. If the mutation occurs upstream, or within a functional domain of the wild-type protein, the physicochemically different +1 protein will lose, either partially or completely, the functional domains of the wild-type protein. At the same time we reasoned that molecular misreading is possible in almost every gene with a GAGAG motif and potentially involved in different pathologies. This was indeed shown to be the case for transcripts associated with Alzheimer's disease and Down's syndrome. Various truncated +1 proteins are present in the neuropathological hallmarks of

GUEST Contd

Alzheimer's disease (viz. neuritic plaques and tangles; see Figure 9.4) and may disturb cellular physiology and initiate neuropathological changes.

VAN LEEUWEN, F., DE KLEIJN, D.P.V., VAN DEN HURK, H., NEUBAUER, A., SONNEMANS, M.A.F.,SLUIJS, J.A., KOYCU, S., RAMDJIELAL, R.D.J., SALEHI, A., MARTENS, G.J.M., GROSVELD, F.G., BURBACH, J.P.H, and HOL, E.M., (1988) Frameshift mutants of β-amyloid precursor protein and ubiquitin-B in Alzheimer's and Down patients. *Science*, **279**, 242–247.

Fred W. van Leeuwen obtained his PhD degree at the Free University of Amsterdam in 1980. Initially, his research concentrated on developing techniques for light and electron microscopical immunocytochemistry. Later, he combined these techniques with molecular biological techniques in a discipline called molecular neuroanatomy. Fred van Leeuwen works at the Netherlands Institute for Brain Research in Amsterdam, an institute of the Royal Netherlands Academy of Arts and Sciences.

3.3 Molecular effects of mutations

In addition to the effects on coding potential illustrated in Table 3.1, DNA mutations can affect gene function and also expression at a variety of levels (Figure 3.1).

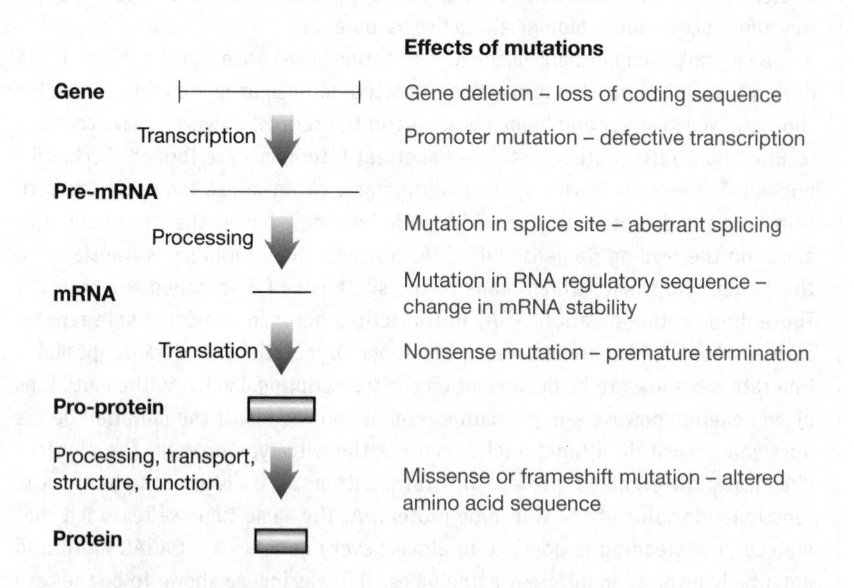

Figure 3.1 DNA mutations may affect gene expression and function at multiple levels.

3.3.1 Aberrant RNA processing in neurones

As indicated in Figure 3.1, mutations in splice sites can lead to RNA splicing defects with potentially crucial functional consequences. Recently, another class of splicing defect has been recognised which does not arise through mutations in the DNA sequence but rather may be caused by a defect in a transcript-specific RNA binding protein (Grabowski, 1998). These defects are interesting for molecular neuroscientists because studies have shown that aberrant RNAs derived from the astroglial glutamate transporter (EAAT2) gene exhibit a regionally specific pattern of expression in the brain, and may contribute to the pathophysiology of a particular neurodegenerative disease, amyotrophic lateral sclerosis (ALS; Lin *et al.*, 1998).

3.4 Mutation analysis in animal models

There has been a long tradition of using the mouse as a mammalian model in genetic studies, but particular rat models have also been extensively studied. One important rat model is the 'Brattleboro' rat (named after the village of Brattleboro in Vermont, USA, where this spontaneous mutant was first discovered) which has a mutation in the vasopressin gene (Valtin *et al.*, 1962; see also Guest Box p. 35).

A large number of spontaneous mouse mutants have also been isolated and maintained as a research resource. A single web site of the Jackson Laboratories in the USA (where many of the available mutants are maintained) serves as a general source of information on mouse mutants (**http://www.jax.org/**). This site also contains a wealth of additional information and useful links; the Mouse Genome Informatics resource is stunning (**http://www.informatics.jax.org/exptools.html**).

In addition to long-established mouse mutants, new mutants are now being generated through major programmes of both directed and random mutagenesis.

3.4.1 Directed mutagenesis

Through the techniques of reverse genetics which now include sophisticated gene targeting approaches, specific genes can be mutated in precisely designed structure/function experiments. These approaches are described in Chapter 4.

Insertional mutations

In transgenic experiments where foreign DNA is incorporated into the mouse genome (see Chapter 4), endogenous genes may be disrupted. If this insertional event is not embryonic lethal, a serendipitous mutant model may be generated. Mutant phenotypes are only recognised following breeding of **hemizygous** transgenic animals to homozygosity because the insertional mutations are generally recessive. An early example of the identification of a transgene insertional mutation was the Purkinje cell degeneration (*pcd*) mouse (Krulewski *et al.*, 1989). In addition to providing an animal model for the further analysis of this particular neurodevelopmental phenotype, transgene insertions can also facilitate the positional cloning (see Box 3.1) of uncharacterised chromosomal loci.

3.4.2 Random mutagenesis

The central importance of animal mutants in molecular neuroscience research is in the provision of a particular phenotype (physiological or behavioural) or model which can be used to advance our understanding of an aspect of brain function, or a specific neurological disease. Although existing mouse models are important in this respect, it has been estimated that the number of known mutant loci is only 1–2% of the total number of mouse genes (Brown and Peters, 1996). This 'phenotype gap' is now being addressed through nationally organised and internationally coordinated programmes of random mutagenesis in which the offspring of mutagenised parent mice are screened for interesting and useful phenotypes. In the UK, a Medical Research Council facility at Harwell is engaged in such a programme (**http://www.mgu.har.mrc.ac.uk/**) and works closely with a mouse genome centre (**http://www.mgc.har.mrc.ac.uk/**) because a key aspect of these initiatives is the rapid identification of novel (mutant) genes. The mutagenesis programme at Harwell (**http://www.mgc.har.mrc.ac.uk/mutabase/description.html**) is currently screening thousands of progeny derived from mutagenised mice each year in order to establish new models of human disease.

The production and analysis of mutant phenotypes with the eventual aim of identifying the genetic basis of the mutation is termed **forward genetics** (from phenotype to genotype; cf. **reverse genetics** – from genotype to phenotype).

Mutagenesis screens

Currently, the favoured animal model for mutagenesis is the mouse. Germline mutations can be generated by a variety of treatments, but the chemical mutagen N-ethyl-N-nitrosourea (ENU) (Table 3.2) is now used widely for several reasons:

- High mutation rate relative to other treatments.
- Generates point mutations which can result in either loss or gain-of-function (see section 3.2). The large deletions caused by other mutagens are often lethal.
- Premeiotic germ cells (spermatogonia) are targeted for mutagenesis, and therefore the first generation of offspring are **non-mosaic**. Other agents target (at least partly) postmeiotic cells.

Table 3.2 Germline mutagens in mice.

Agent	*Target*	*Mutation rate per locus (10^{-5})*	*Predominant mutation*
X-rays	Spermatogonia	50	Small deletions
	Postmeiotic germ cells	33	Deletions, translocations
	Oocytes	19	Deletions, translocations
ENU	Spermatogonia	150	Intragenic point mutations
Chlorambucil	Postmeiotic germ cells	127	Deletions, translocations
None		0.5–1.0	

Adapted from Rinchik, E.M. Chemical mutagenesis and fine-structure functional analysis of the mouse genome. *Trends in Genetics*, 7, 15–21. Copyright 1991, with permission from Elsevier Science.

Mutation rates are with respect to optimal dose regimens. ENU, *N*-ethyl-*N*-nitrosourea.

The ENU mutagenesis strategy is outlined in Figure 3.2. In this approach, the offspring of mutagenised males can be screened directly for **dominant** mutations, whereas a more extensive, three-generation breeding programme is required to recover **recessive** mutations. Given the mutation frequency of ENU (Table 3.2), it has been estimated that the screening of 3000 offspring in a dominant screen represents genome-wide **saturation mutagenesis** where all genetic loci have been mutated at least once (Takahashi *et al.*, 1994). Thus, a researcher can be fairly confident that their 'gene' and associated mutation phenotype will be detected by following this programme. In practice, if one is fortunate, such an extensive programme of breeding and screening may not be necessary – for example, considerably fewer offspring were screened before a desired 'clock' mutant (daily behavioural rhythm) mouse was obtained using this strategy (see section 8.3.2).

In addition to choosing an appropriate mutagen, the success of a mutagenesis screen approach is also dependent upon the **screening procedure** which must be:

- easy to perform, if not automated (for example, behavioural rhythm phenotypes are screened by housing mice in a wheel-running chamber where rhythms of locomotor activity are monitored through computer links; section 8.3.2).
- unambiguous (the screen must be designed such that only the required phenotype is detected – for example, mice with a general loss of locomotor control may also exhibit aberrant rhythms of daily activity that are unrelated to 'clock' gene function).

The programmes of random mutagenesis which have been established to provide a variety of phenotypes for the research community (see section

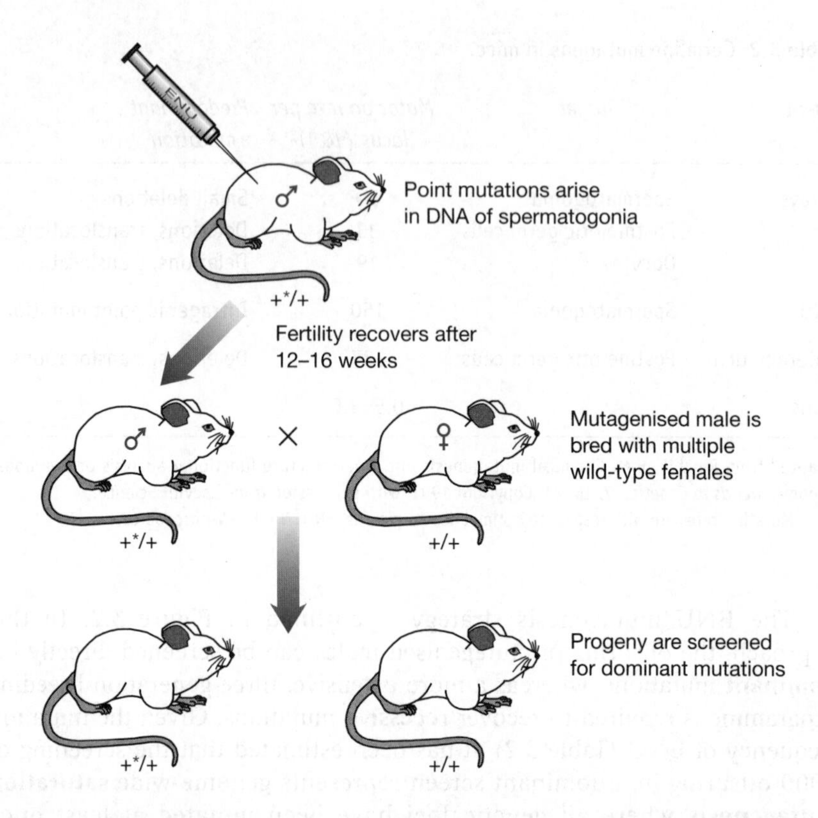
ENU
Point mutations arise
in DNA of spermatogonia
+*/+
Fertility recovers after
12–16 weeks
Mutagenised male is
bred with multiple
wild-type females
+*/+
+/+
Progeny are screened
for dominant mutations
+*/+
+/+

Figure 3.2 ENU mutagenesis screen. The chemical mutagen ENU (*N*-ethyl-*N*-nitrosourea) is injected into a single male mouse, and multiple offspring are screened for the desired phenotype. Note that in this screen the offspring are heterozygous (asterisk) for particular mutations, and therefore only dominant phenotypes will be observed in the screen.

3.4.2) have developed integrated screening procedures which are designed to identify multiple different phenotypes. An example is the SHIRPA procedure: (**http://www.mgc.har.mrc.ac.uk/mutabase/shirpa_summary.html**) which has been formulated for comprehensive phenotype assessment (Table 3.3; Rogers *et al.*, 1997).

Following a successful mutagenesis screen, a novel animal model is established that may be bred extensively and used for behavioural, physiological and pharmacological studies. However, the ultimate aim is the identification of the mutated gene.

3.5 Identification of mutant genes

Gene mutations, which either form the basis of inherited human diseases, or have been introduced randomly into animal models (see section 3.4.2), remain equivocal until the mutated gene is cloned, and the DNA sequence is compared with the wild-type sequence.

Table 3.3 SHIRPA protocol for comprehensive phenotype analysis.

Primary screen
Behavioural observation profile (records overt behavioural abnormalities)

Secondary screen
Locomotor activity (automated recording of activity over 24 hours; see section 3.4.2)
Food and water intake
Balance and coordination
Analgesia (standard hot plate test)
Histology (includes X-ray analysis)
Biochemistry (e.g. analysis of blood glucose and electrolytes)

Tertiary screen
Anxiety (open field activity; see Figure 10.2)
Learning and memory (e.g. Morris watermaze; see section 8.2.1)
Prepulse inhibition (indicates schizophrenia-like deficits; see section 10.2.6 and Dulawa *et al.*, 1997)
Electroencephalography
Nerve conduction
Magnetic resonance imaging

The name SHIRPA is an acronym partly based on the institutions where the protocol was devised: **S**mithKline Beecham Pharmaceuticals; **H**arwell, MRC Mouse Genome Centre and Mammalian Genetics Unit; **I**mperial College School of Medicine at St Mary's, **R**oyal London Hospital, St Bartholomew's and the Royal London School of Medicine; **P**henotype **A**ssessment.

Comprehensive details of the tests can be found at: **http://www.mgc.har.mrc.ac.uk/mutabase/shirpa_summary.html.**

It should be noted that the procedure can be tailored to identify particular behavioural abnormalities. Also individual procedures within SHIRPA should be carried out in order of increasing physical manipulation, such that the least stressful tests are carried out first.

In some instances, it is possible to make an informed guess (based on the function of known genes) concerning the identity of the mutated gene that underlies a particular phenotype, and it is then a relatively simple matter to obtain the '**candidate**' gene sequence from the affected individual/mutant (see cloning strategies in Chapter 1). However, it is much more likely that the identity of the mutated gene is quite obscure. For this reason, a new molecular genetic technique termed **positional cloning** has been developed (Box 3.1). A recent application of this approach is described in section 8.3.2 (the *clock* gene).

One of the best illustrations of research on the mouse (involving positional cloning) leading to a major discovery in human genetics is described in the Guest Box on p. 43.

BOX 3.2: POSITIONAL CLONING

Positional cloning describes a series of associated techniques that are used to 'home in' on an unknown gene. This approach involves moving from a starting point of low resolution (...'we have a mutant mouse') through a series of mapping procedures that increase the level of resolution from the genetic to the physical (see Chapter 2), and eventually to the individual gene. Positional cloning generally proceeds through the following steps, but it should be recognised that current gene, and genome, sequencing efforts are rapidly changing the ground rules of this research field (see Chapter 2).

1. Establish an initial chromosomal map position by **linkage analysis** using **polymorphic markers** such as **SSLPs** to determine the pattern of inheritance (section 2.2.1). This procedure identifies a 'critical region' in which either the mutated gene (mouse) or disease locus (human) resides. **Candidate genes** may be apparent at this early stage (section 3.5), or at later stages in the genetic mapping procedure.
2. Obtain a high-resolution linkage map by selecting flanking markers that lie within a few cM (section 2.2.1) of the mutation. Eventually, high-resolution mapping can provide a map position within a region of less than 1 cM, which is similar (in physical units) to the size of **YAC** clones (section 1.3.1).
3. The transition from genetic to physical mapping is attained by using the linked markers as probes to isolate DNA clones (e.g. **YAC**) which contain the marker sequences.
4. A mutually overlapping set of clones called a **contig** (section 2.2.2; so-called because adjacent clones are contiguous) which spans the entire region between the flanking markers is obtained.
5. At this stage DNA clones may be used as **transgenes** in transgenic mice to **rescue** (section 8.3.2) the mutation – proving that the clone contains the wild-type equivalent of the mutated gene. However, this does not preclude the identification and sequencing of **transcription units** (genes) within the clones.
6. The final route to gene identification is partly dictated by the (predicted) nature of the gene, which may, for example, be selectively expressed in a particular tissue. In this case, a tissue-specific cDNA library (section 1.2.1) can be probed with contig clones. In practice, a variety of different approaches may be used at this stage of positional cloning (see King *et al.*, 1997). One approach, successfully used by King *et al.*, is simply to obtain (thousands of) subclones of the large contig clones, and sequence these – **shotgun sequencing**. Sequence similarity searches (Box 1.3) may then be used to identify candidate expressed sequences. Other approaches include exon-amplification (see Takahashi *et al.*, 1994)
7. Once the mutated gene is identified, a full-length clone must be obtained and the sequence compared with the wild-type.

GUEST BOX BY KAREN B. AVRAHAM

Tales/tails from the ear – the genetics of hearing loss

The Sanskrit word for mice, mush, means 'to steal', and appropriately is the name given to mice due to their reputation in ancient times and beyond. However, mice have vindicated themselves in the world of genomics by providing the key to the discovery of several human disease genes. The study of the auditory system has been no exception. The physiology and morphology of the inner ear are remarkably similar between mice and humans, and the large number of mutant mice with inner ear defects have provided an invaluable resource. Hearing loss affects as many as 10% of the human population, but the isolation of genes for deafness has only recently begun, lagging behind other disciplines, such as blindness and cancer. Inherited disease genes are identified using linkage analysis and positional cloning (Box 3.2). In an effort to isolate the defective gene in Family H, a large Israeli family with over 50 members exhibiting progressive hearing loss, such a positional cloning approach was taken. Ten months after collecting blood from 23 family members, in a collaboration spanning three coastlines, the Mediterranean, the Pacific and the Atlantic, linkage analysis defined the critical disease region to 25 cM on chromosome 5. In the human genome, this equals roughly 25 million bp. Imagine how long it would take to sift through this enormous region – not a project a student should undertake to ensure a successful thesis! However, hard work, in conjunction with serendipity, can go far. The defective **locus** in Family H just happened to lie very close to the first deafness locus mapped, DFNA1. Initially, DNA from affected Family H members was being screened for mutations in diaphanous, the protein encoding the DFNA1 locus. None was found and soon thereafter, critical recombinations in two Family H members revealed that a new locus had been identified, named DFNS15. The time for the **candidate** gene approach to cloning a gene had arrived, and the 25 cM region was investigated for previously identified genes. Human and mouse chromosomes share regions of homology, and human 5q31 was no exception, sharing homology with mouse chromosome 18 containing the transcription factor Pou4F3. The fit was perfect – Pou4F3 is specifically and uniquely expressed in the hair cells, the sensory cells of the inner ear, and when removed from the mouse genome by gene targeted mutagenesis, leads to profound deafness in mice. The human homologue, POU4F3, was amplified by PCR and sequenced. A mutation was found in POU4F3 only in hearing impaired individuals of Family H! The pace at which events took place was daunting, with three laboratories (along with help from others) literally coordinating their efforts around the clock. While the Israeli contingent was sleeping, the East coast scientists were conducting experiments and the West coast colleagues were thinking of waking up. And within weeks, the continued efforts of all groups involved led to the publication of the findings in the journal *Science*.

GUEST BOX Contd

Vahava, O., Morell, R., Lynch, E.D., Weiss, S., Kagan, M.E., Ahituy, N., Morrow, J.E., Lee, M.K., Skvorak, A.B., Morton, C.C., Blumenfejd, A., Frydman, M., Friedman, T.B., King, M.-C. and Avraham, K.B. (1998) Mutation in transcription factor *POU4F3* associated with inherited progressive hearing loss in humans. *Science*, **279**, 1950–1954.

Karen Avraham has travelled extensively from her native Canada and USA, obtaining a BA from Washington University in the USA and PhD from the prestigious Weizmann Institute in Israel. Following a postdoctoral fellowship at the National Cancer Institute in Maryland, she is currently an assistant professor in the Sackler School of Medicine, at Tel Aviv University.

3.6 Summary

- Gene mutations may be either functional (causing either a loss or gain of function) or silent.
- DNA mutations are of two basic types – length and point.
- Molecular misreading is a new class of mutation in which mutant RNAs are transcribed from normal (wild-type) DNA.
- Aberrant RNA processing is another new class of mutation recently observed in neurones.
- Mutation analysis in experimental animals is a key approach for understanding gene function – both directed and random mutagenesis (e.g. ENU) are used.
- The study of mouse mutants is contributing to the characterisation of inherited diseases in humans, for example in hearing loss.
- Genome-wide mutagenesis screens are in progress using integrated screening procedures.
- Positional cloning techniques are used to 'home in' on (novel) mutant genes.

Further reading

EVANS, M.J., CARLTON, M.B.L. and RUSS, A.P. (1997) Gene trapping and functional genomics. *Trends in Genetics*, **13**, 370–374.

Discussion of an alternative approach to random mutagenesis in mice.

ZAMBROWICZ, B.P., FRIEDRICH, G.A., BUXTON, E.C., LILLEBERG, S.L., PERSON, C. and SANDS, A.T. (1998) Disruption and sequence identification of 2,000 genes in mouse embryonic stem cells. *Nature*, **392**, 608–611.

*As the genome projects progress, and all mammalian genes are identified, the emphasis of genomics research will switch to addressing gene function. Analysis of induced mutations will be crucial. However, the process of making gene knockouts by homologous recombination in ES cells is difficult and time consuming (see Chapter 4). It would be much better if, having identified a gene sequence, one could go to a library and take the corresponding knocked-out ES clone down from the shelf. The Omnibank system, developed by Lexicon Genetic Incorporated (**www.lexgen.com**) does just that.*

CHAPTER FOUR

Germline transgenesis

Key topics

- Why is transgenesis so important?
- The germline cycle
- Transgenic methods
    - Microinjection of fertilised one-cell eggs
    - Retroviral infection of embryos
    - Embryonal stem cells and gene knockouts
    - Nuclear transfer and cloning
    - Manipulation of male germ cells
- Limitations and complications of germline transgenesis
- Cell-specific and inducible transgenes
- Cell-specific and inducible gene knockouts
- Alternatives to gene knockouts
    - Antisense RNA
    - Dominant negatives
    - Cell ablation
    - Recombinant antibodies
- Genetic background

4.1 Introduction

Transgenesis is the key technique in contemporary molecular neuroscience. If we are to understand the complexities of neuronal function, we must study the role and regulation of genes in the context of the complexities of intact brain. Further, the functions of all of the new genes identified as part of the various **genome projects** (see Chapter 2) must be elucidated in physiological terms – a concept known as **'functional genomics'**.

In the past, neuroscientists have addressed these issues by introducing pharmacological agents into the brain, for example, agonists or antagonists that affect particular pathways. However, even if such reagents are available for the gene of interest, this approach has its problems, for example, lack of cell specificity, the need for surgical intervention, toxicity and non-specific drug effects. Transgenesis, in contrast, enables brain gene function to be studied in the brain, with precise spatial and temporal specificity.

The term transgenesis is used very broadly to describe the introduction of cloned DNA into any living cell. Over the past 10–15 years, profoundly important new techniques have been developed that enable new genes to be introduced into whole mammalian organisms. The resulting animals are called transgenic animals, defined as any animal into which cloned genetic material has been transferred.

Two types of transgenesis can be defined:

- **Germline** – the transgene is present in the chromosomes of every cell in the organism, and is transmitted to subsequent generations. Germline transgenesis is discussed in this chapter. More information about germline transgenics, and a comprehensive database on different types of transgenic animal, can be found at the TBASE site:
  http://www.jax.org/tbase.
  A good transgenics links page is to be found at:
  http://www.rodentia.com/wmc/.
- **Somatic** – the transgene is directed to specific somatic cells in an individual organism, and is not transmitted to subsequent generations (see Chapter 5).

4.2 Why transgenesis?

Making and analysing transgenic animals is time consuming and expensive. So why bother?

4.2.1 Negative reasons

All other methods for studying the regulation and function of mammalian genes are deficient.

Limitations of classical mammalian genetics

Before the current revolution in transgenic research, the only method available to study the regulation and function of mammalian genes within the context of the whole organism was to utilise mutants that arose spontaneously in nature. The study of such mutations has been termed **'forward genetics'** – from phenotype to gene (see section 8.3.2). Disadvantages of this approach are that:

classically use of forward genetics (spontaneous mutations)

1. mutations arise opportunistically and those affecting particular systems are difficult to screen for and identify;

2. the **congenic** breeding experiments required to isolate the genetic defect on a different genomic background are difficult to perform; and
3. the molecular characterisation of the mutation is technically difficult.

Limitations of *in vitro* systems

A number of techniques are available that enable characterised, cloned gene sequences to be transfected into cultured mammalian cells. Following transfection, the DNA may be carried as an extrachromosomal plasmid (transient transfection), or selection can be performed to isolate clones in which integration into the cellular chromosomes has occurred (stable transfection). Evaluation of gene expression and function in cell lines is relatively straightforward. Disadvantages of this approach are that:

- cell lines arise opportunistically in nature, usually as a result of the oncogenic process (see **http://www.atcc.org/** for a comprehensive database of mammalian cell lines). Often, no appropriate cell lines exist for the cell type of interest.
- no cell line can ever model the plasticity evoked by the panoply of developmental and physiological stimuli to which a mammalian gene is normally exposed within the physiological integrity of the whole organism.

4.2.2 Positive reasons

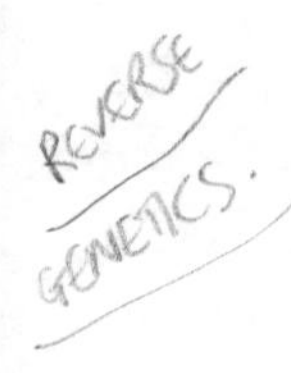

Transgenic animal systems combine the virtues of cell culture and congenic breeding strategies while avoiding the negative aspects of each system. In a transgenic animal, a defined gene sequence is studied within the physiological and developmental integrity of the whole animal, without any prior knowledge of its regulation or function. This has enabled the development of the concept of **reverse genetics** – from gene to phenotype.

4.3 Altering the germline

Mammalian reverse genetics demands that a defined genetic change is introduced into every cell of the organism, and is transmitted to subsequent generations. These ends are met by altering the germline (see Box 4.1). Five techniques have been developed:

1. Direct microinjection into the pronuclei of fertilised one-cell eggs.
2. *In vitro* infection of embryos with retroviruses.
3. *In vitro* manipulation of pluripotential embryonic stem (ES) cells.
4. Nuclear transfer.
5. *In vitro* manipulation of male stem cells.

BOX 4.1: THE GERMLINE CYCLE

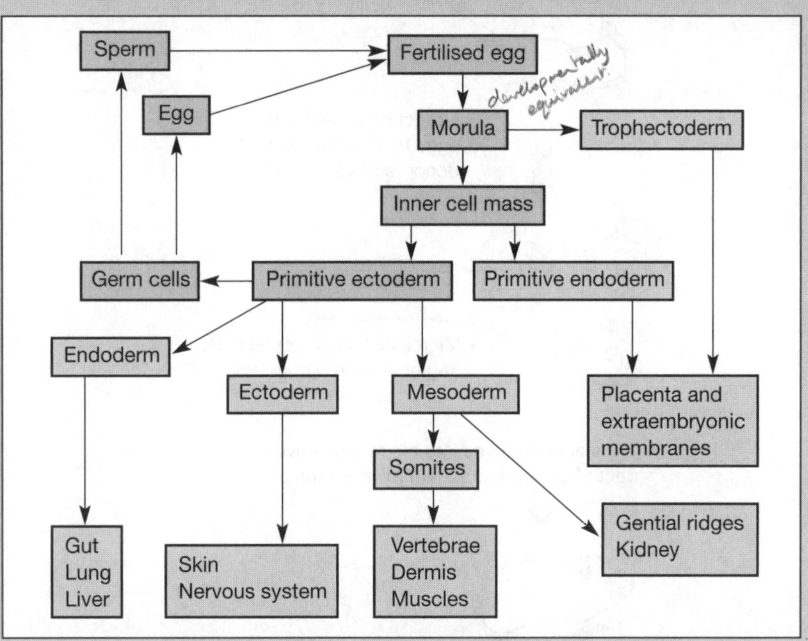

A mammalian organism consists of a huge number of different cells each organised into distinct tissues and organs that have particular functions. All of these cells are derived from the successive divisions of the fertilised one-cell egg, the product of the union between the egg and a successful individual sperm. The fertilised one-cell egg gives rise, firstly, to a ball of cells called the morula, all the cells of which are developmentally equivalent – they can all give rise to all of the other cells of the embryo. By the blastocyst stage, the first differentiation step has taken place. The blastocyst consists of two cell types – the trophectoderm, which gives rise to tissues that surround, nurture and protect the embryo, such as the placenta, and the inner cell mass (ICM). The ICM then differentiates to give rise to the primitive endoderm, which engenders yet more extraembryonic tissues, and the primitive ectoderm, which gives rise to all cells of the embryo proper, including the germ cells. The germ cells give rise to the egg and the sperm, which once united through fertilisation, set the cycle, the germline cycle, off again. If a genetic change can be introduced into this cycle, all else follows – all subsequent cells carry change, and all subsequent generations carry change.

4.3.1 Microinjection

The process of making transgenic mice by the injection of a cloned DNA fragment into fertilised one-cell eggs is illustrated in Figure 4.1.

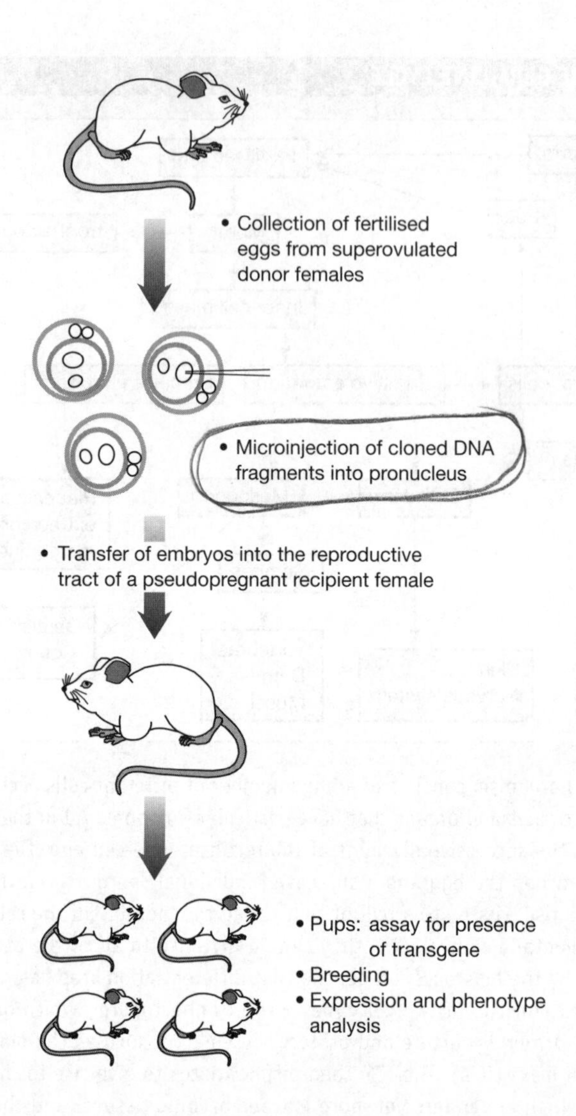
• Collection of fertilised eggs from superovulated donor females
• Microinjection of cloned DNA fragments into pronucleus
• Transfer of embryos into the reproductive tract of a pseudopregnant recipient female
• Pups: assay for presence of transgene
• Breeding
• Expression and phenotype analysis

Figure 4.1 Summary of the process of making transgenic mice by the microinjection of fertilised one-cell eggs with cloned DNA fragments. One-cell fertilised eggs are harvested from donor females that have usually been hormonally stimulated to produce a huge number of eggs before mating. Natural matings can be used, but these will only give 10 fertilised one-cell eggs per donor female. Superovulation of immature females increases the yield of fertilised one-cell eggs to between 30 and 50 per donor female. Superovulation involves the injection of a dose of pregnant mare's serum, to mimic follicle stimulating hormone (FSH), followed a couple of days later by an injection of human chorionic gonadotrophin (hCG), to mimic luteinising hormone. Ovulation takes place 10–13 hours after the hCG injection. The following morning, donor females are killed and the eggs are harvested from the oviduct. Cloned DNA fragments are then directly introduced into one of the two pronuclei using a fine injection needle with a 1 mm opening. Approximately 1–2 pl of a solution of 1–2 mg/ml of DNA is introduced into the nucleus, which corresponds to a few hundred molecules of DNA per injection. Surviving eggs are transferred to the natural environment provided by a pseudopregnant recipient female. A pseudopregnant recipient is an oestrus female that has been mated with a sterile or vasectomised male. The animal carries only unfertilised eggs, but is physiologically prepared to carry implanted eggs through pregnancy. Pups are born 20–21 days later; of these, 10–50% are transgenic and are identified by genome analysis.

The process of microinjection results in the integration of the transgene into the host chromosomes. This is thought to take place at the one-cell stage. Thus, in most cases, the animal that develops from the injected egg will contain the transgene in all of its cells. As this includes the germline, the transgene will be passed on to subsequent generations. Occasionally, integration will not occur until the 2-, 4- or 8- cell stages of development. In such a case, only a proportion of the cells of the resulting animal will be transgenic. The animal will be **chimeric,** or mosaic. However, if the germline has been colonised, even at a low efficiency, then a pure transgenic mouse can be derived at the next generation.

Integration is an additive process; the host genome gains new information. The host genome is unchanged, except at the locus of integration, where deletions may occur, possibly interrupting endogenous genes. Integration takes place through **non-homologous recombination** of the transgene into the host chromosomes, and there is no specificity with respect to either the host or the transgene DNA. There is usually only a single integration site per nucleus, but that integration site might contain between 1 and 1000 tandemly arranged copies of the transgene. The integration process is essentially random; the experimenter has no control over the site of integration, nor the copy number of the transgene. A transgene is expressed with a spatial and temporal pattern that is a function of the ***cis*-acting elements** that it contains. The structure of a microinjection transgene depends entirely on the aims of a particular experiment. However, certain general rules should be considered when designing a construct for eventual introduction into a transgenic organism (Box 4.2). A transgene should operate like any other gene in the targeted cells of an organism, and hence, structural elements must be appropriately recognised by the transcriptional, post-transcriptional and translational machinery of the host (Box 4.3).

BOX 4.2: TRANSGENE DESIGN

1. The process of making transgenic animals is both lengthy and laborious. There is a need to consider carefully whether further cloning procedures would give the experiment a greater chance of success. Having built a transgene, the experimenter might wish to test whether it is capable of acting as a functional expression unit in transfected culture cells before proceeding to the transgenic animal.
2. The design of the transgene should take into account strategies for differentiating between the transgene and the endogenous gene, at the level of genomic and expression analysis. Sequence differences between species can be exploited or **reporters** can be included (see Chapter 6).
3. It is usual to inject DNA fragments. Linear DNA has been shown to integrate five-fold more efficiently than supercoiled DNA, but the structure of the fragment ends created by different restriction enzymes has no effect.

BOX 4.2: Contd

4. All vector sequences should be removed before injecting cloned eukaryotic genes in order to maximise the quality, quantity and reproducibility of transgene expression. The presence of contiguous vector-derived DNA sequences in a fragment of injected DNA has been shown to inhibit the expression of mammalian transcription units in mice.
5. Position effects, based upon the chromosomal location of the integrated transgene, can have profound effects on expression, resulting in repression if integrated in a region of transcriptionally inactive chromosome, or ectopic activation if integrated adjacent to a powerful enhancer element. It is therefore necessary to demonstrate the same pattern of transgene expression, or the same functional consequence of transgene expression, in at least two independently derived transgenic animal lines.

BOX 4.3: TRANSGENE EXPRESSION

Transcription (http://www.mc.vanderbilt.edu/gcrc/gene/transg.htm)

1. What is known about the promoter and enhancer sequences of the gene of interest?
2. Have studies on the transfection of the gene into cultured cells identified any elements that should be included or eliminated from the transgene?
3. If little is known about the transcriptional regulatory elements, then one should aim to build as big a transgene as possible, even resorting to the use of a cosmid (30–40 kb), bacterial artificial chromosome (BAC, 100–150 kb) or yeast artificial chromosome (YAC, up to 250 kb) clones.
4. If the transgene is from a different species than the host, are species differences in expression likely to complicate the analysis of expression?
5. Do not neglect introns or sequences beyond the 3' end of the last exon. Non-coding sequences can contain important regulatory elements, and regions many kb upstream or downstream of the start of transcription can contribute to the proper control of transcription initiation and termination. Introns have been shown to greatly improve the efficiency of overall transgene activity, possibly by directing appropriate packaging of the transgene transcription unit into chromatin.
6. Consider the position of regulatory elements within a transgene. Although enhancers are orientation- and position-independent, other signals, such as the TATA box, should be correctly positioned with respect to a functional transcription start site.

Post-transcriptional processes

1. The first and the last exon of the transgene can be any length. However, the size of internal exons is constrained, possibly by the machinery that recog-

BOX 4.3: Contd

nises exon/intron boundaries and engenders splicing. The construction of internal exons of greater than 300 bp by, for example, the insertion of a reporter element, may result in poor splicing efficiency and low expression of the transgene overall.

2. The proper location of signals for polyadenylation is important, not only to engender proper synthesis of the poly (A) tract, but also to facilitate appropriate transcription termination and mRNA 3' end formation.

Translation

1. A reporter gene, which is best located in the first or last exon, should possess a suitably located, functional translation initiation codon (AUG) within the context of an efficient **Kozak** translational start sequence. Care should be taken to ensure that no functional initiation codons are present upstream of the one that should be utilised.
2. Alternatively, the reporter can be synthesised as an in-frame fusion with another coding sequence.

Post translational processes

1. The protein product of the transgene may be destined for a particular subcellular location. Does the transgene contain the appropriate signals, such as those that mediate, for example, processing, secretion and nuclear localisation?
2. The transgene protein product may require post-translational modification for activity, for example, phosphorylation or glycosylation. Are the signals that engender these processes present and/or accessible?

4.3.2 Retroviruses

Retroviruses have evolved to deliver their genomes efficiently into host cells. As such, they make ideal gene transfer vectors if they can be suitably modified **(http://www-micro.msb.le.ac.uk/335/Retroviruses.html).** Retroviruses (see Box 4.4) carry an RNA genome that is copied into DNA once in the infected cell, and then integrates into the host's chromosomes. Once integrated, the viral genome is expressed, replicates along with the host DNA and is transmitted to all daughter cells. The retroviral genome consists of little more than the genes essential for viral replication. These can be replaced with the transgene. Infectious recombinant retrovirus particles are made by introducing recombinant DNA into a packaging cell line that makes the viral proteins required for capsid production and virion maturation *in trans*. To make transgenic animals, morulae are dissected into culture along with packaging cells that are producing the recombinant virus. Infected embryos are returned to the natural environment provided by a pseudopregnant recipient surrogate mother, who will give birth to pups bearing the recombinant provirus.

Advantages of the use of recombinant retroviruses to make transgenics are:

- efficient integration;
- single-copy integration; and
- high infectivity.

Disadvantages of the use of recombinant retroviruses to make transgenics are:

- Recombinant retroviruses can only accommodate a small 6–9 kb insert.
- Recombinant retroviruses are potentially unstable, and could represent a potential biological hazard for two reasons:
  (i) the recombinant, replication-deficient virus could recombine with endogenous proviral sequences, resulting in the production of infectious virus;
  (ii) long terminal repeats (LTRs) are powerful promoters of transcription. These can interfere with transgene expression, and harmful activating mutations could result from integration of the proviral LTRs next to a cellular gene such as a proto-oncogene.

  These problems can be overcome by the use of suicide or self-inactivating (SIN) viruses. In vectors that give rise to SIN viruses, the 3' LTR contains a deletion that renders it non-functional. The intact 5' LTR permits replication and the production of infectious virions in a packaging cell. However, once inside a target cell, the 3' LTR serves as the template for construction of both the 5' and the 3' LTR of the provirus. Thus both the 5' and the 3' LTRs of the provirus are non-functional; the provirus cannot be rescued, there can be no further transmission, and the LTR cannot interfere with expression of transgene or neighbouring cellular genes.
- The high infectivity of retroviral particles means that an embryo is infected many times. Each cell in each embryo contains many different integrated viruses at many different chromosomal positions. The resulting transgenic pups are thus highly mosaic. Different tissues are derived from different early embryo precursors with different patterns of virus integration. Considerable outbreeding is needed to produce pure stable lines containing a limited number of proviral integration events.

Insertional mutagenesis

When a provirus inserts into a chromosome, it may integrate into a gene, thus disrupting its normal function (also see section 3.4.1). The resulting animal will thus be heterozygous, having one normal and one mutated allele at this locus. If two such animals are mated, the mutation is revealed

BOX 4.4: RETROVIRUSES

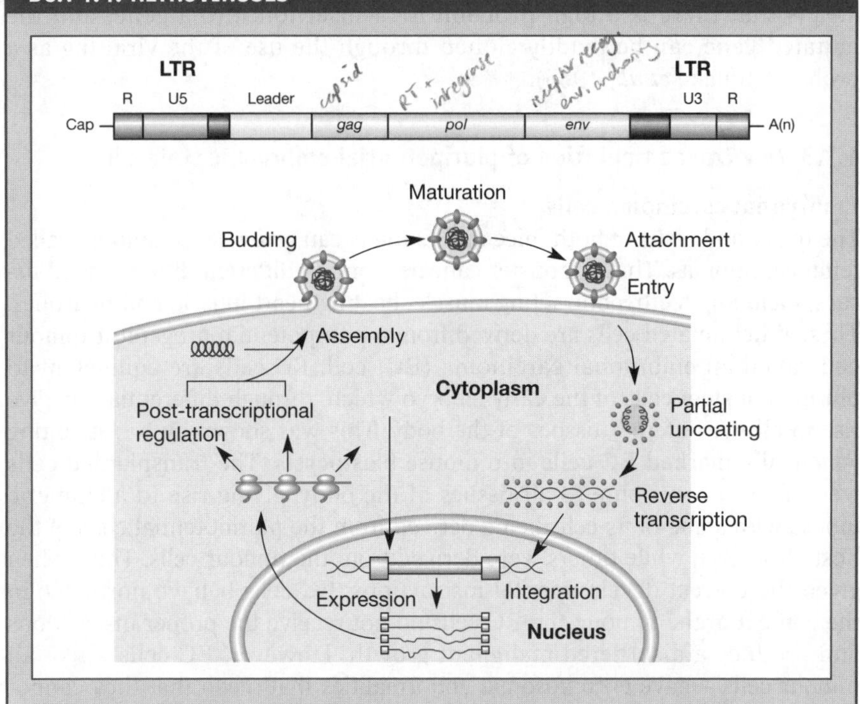

The retrovirus genome consists of two identical positive (sense) single-stranded RNA molecules ranging in size from 3.5 to 10 kb. The retrovirus virion contains a protein capsid that is lipid encapsulated. The viral genome is encased within the capsid, along with the virally encoded integrase and reverse transcriptase proteins. The gag region encodes genes which comprise the capsid proteins; the pol region encodes the reverse transcriptase (RT) and integrase proteins; and the env region encodes the proteins needed for receptor recognition and envelope anchoring. The long terminal repeat (LTR) regions found at each end of the genome play an important role in controlling viral replication and transcription. Following infection and uncoating, RT synthesises a DNA copy of the viral RNA template primed by a host tRNA, and viral RT. Conversion to double-stranded DNA is a complex process mediated by host enzymes. Importantly, the 3' LTR acts as template for both of the LTRs of the resulting linear double-stranded DNA. This circularises and integrates into the host chromosomes through the action of the viral integrase, giving rise to the provirus. Integration is random with respect to the host genome, but is specific with respect to the viral DNA. The provirus consists of a single copy of the viral genome flanked by LTRs which mediate transcription. Retroviruses can be made to infect an early embryo. The integrated viral genome will be transmitted to the cells of the adult organism, and to subsequent generations through the germline. A recombinant retroviral vector is made by replacing the *gag*, *pol* and *env* genes with the transgene. The vectors thus contain the transgene flanked by LTRs. Infectious recombinant retrovirus particles are made by introducing recombinant DNA into a packaging cell line that makes the viral proteins required for capsid production and virion maturation, the gag, pol and env proteins, in *trans*.

gag, pol + env replaced by transgene.

in the homozygous offspring. The highly efficient infectivity of retroviruses means that there is a high probability of insertion into a gene, and the mutated gene can be readily cloned through the use of the viral tag as a probe (Schnieke *et al.*, 1983).

4.3.3 *In vitro* manipulation of pluripotential embryonic stem cells

Embryonal carcinoma cells

The ovary and testis of both mice and humans can give rise to tumours called teratocarcinomas. These 'monster cancers' contain different differentiated tissues, including teeth, contracting muscle, beating heart muscle and neurones. These differentiated cells are derived from a pluripotential progenitor tumour cell called an embryonal carcinoma (EC) cell. EC cells are equivalent to pluripotent stem cells of the early embryo which, through differentiation, give rise to all the different tissues of the body. This was shown by transplanting genetically marked EC cells into mouse blastocysts. The transplanted cells were able to contribute to all tissues of the body, giving rise to a chimeric mouse with some of its cells being derived from the pluripotential cells of the host blastocyst, while others were derived from the tumour cells. Thus, when given the correct developmental instructions, EC cells behave normally; in the context of the tumour, the EC cells do not receive the proper instructions and produce a disordered malignant growth. However, EC cells - like all tumour cells - have chromosomal abnormalities that mean that they cannot contribute to the germline of chimeric mice. That is, an EC cell cannot pass on its genetic information to a subsequent generation.

Embryonal stem cells

Embryonal stem (ES) cells are pluripotential stem cell lines derived from the primitive ectoderm of the mouse blastocyst (Brook and Gardner, 1997). ES cells have the same characteristics as EC cells, but have a normal karyotype and are thus able to contribute to the germline. They are able to give rise to eggs and sperm, and thus their genetic information can be passed on to subsequent generations

ES cells as a tool for mammalian genome engineering

As with any mammalian cell in culture, DNA can be introduced into ES cells. ES cells can be genetically altered and these altered cells can be introduced into a blastocyst where they will contribute to the normal development of a chimeric mouse. If the germline of the mouse has been colonised by the ES cell descendants, then, at the next generation, a pure line of heterozygous genetically altered mice can be derived.

ES cells can be altered in two ways:

1. **Non-homologous recombination.** This is by far the most common mode of integration of exogenous DNA into the chromosomes of mammalian cells. Integration is random, and there is no specificity

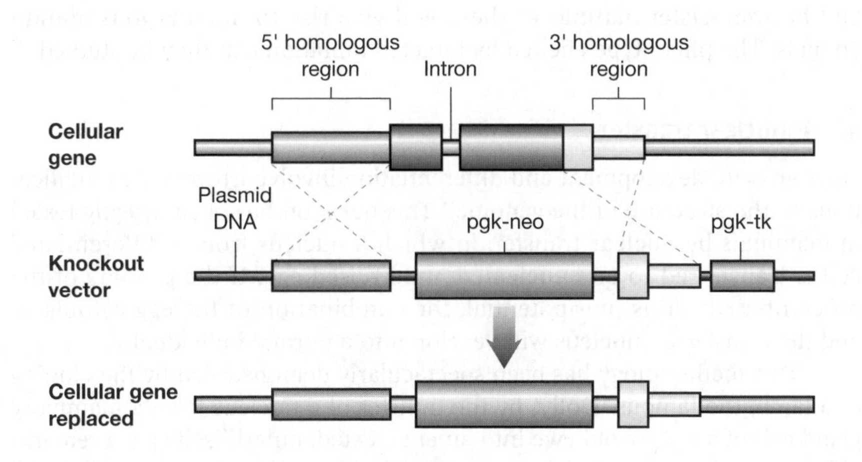

Figure 4.2 Gene targeting in ES cells and identification of homologous recombinants by positive-negative selection. The targeting vector contains two selectable markers, one that can be selected for, one that can be selected against. The positive selection is the G418 resistance gene (an aminoglycoside phosphotransferase also known as the neomycin resistance gene, neo^r). Cells will die in the presence of the drug G418 unless neo^r is being expressed. Expression of neo^r is mediated by the pgk promoter. Negative selection is mediated by the herpes simplex (HSV) thymidine kinase (TK) gene under the control of the pgk promoter. Cells expressing HSV-TK will die in the presence of the nucleoside analogue ganciclovir through inhibition of DNA elongation. Cells that have acquired the neo^r gene by non-homologous recombination will usually also have the pgk-tk gene, and will die in the presence of gangciclovir. Cells that have acquired the neo^r gene by homologous recombination will have lost the pgk-tk gene and will survive in the presence of gangclovir.

with respect to either the host or input DNA. The host gains new information, with the rest of the genome remaining essentially unchanged, except for deletions or other rearrangements around the integration site.

2. **Homologous recombination.** Homologous recombination involves an exchange of information between the input DNA and homologous host sequences, mediated by Watson–Crick base-pairing and host recombination enzymes. Homologous events are very rare compared with non-homologous events. Screening systems (for example, positive/negative selection; see Figure 4.2) have been developed that enable clones bearing these rare homologous events to be isolated.

Using appropriately designed targeting vectors, specific mutations (for example, null mutants, known as **knockouts**) can be introduced into target genes in ES cells. The mutated ES clone can then be introduced into a blastocyst, where it will colonise the inner cell mass and contribute to the development of a chimera. Breeding of a germline chimera with a wild-type mate results in the generation of a pure line heterozygous for the mutation, and brother–sister matings of these will give rise to homozygous mutant animals. The phenotype elicited by targeted mutation can then be studied.

4.3.4 Nuclear transfer

Does growth, development and differentiation involve irreversible modifications to the structure of the genome? This question has been directly tested in mammals by nuclear transfer, in which a nucleus from a differentiated cell is transferred to an enucleated, unfertilised egg. If the genome of the differentiated cell is pluripotential, the combination of the egg cytoplasm and the transferred nucleus will develop into a normal individual.

This methodology has been spectacularly demonstrated by the cloning of a lamb, the famous 'Dolly', by the transfer of a nucleus from a mammary gland cell of a 6-year-old ewe into an enucleated, unfertilised egg; a remarkable, but controversial, demonstration that a nucleus from an adult cell might be able to engender normal development (Wilmut *et al.*, 1997). The genome of the parent cell had not gone through any irreversible changes, and was still pluripotent. Differentiation is thus brought about by systematic and sequential changes in gene expression mediated by interactions between the nucleus and the changing cytoplasmic environment.

Nuclear transfer provides a novel way of genetically manipulating higher organisms.The mouse is the only species for which ES cells have been derived. Knockouts have thus not been possible in other species. However, rather than relying on ES cell technology, it is now possible to manipulate genetically a mature cell in culture, and then transfer the nucleus into an oocyte recipient in order to introduce the change into the whole organism. Already, transgenic sheep carrying the human clotting factor IX gene have been made by transferring nuclei carrying the transgene from transformed sheep fibroblasts into recipient oocytes (Schnieke *et al.*, 1997).

4.3.5 Manipulation of male stem cells

Diploid stem cell spermatogonia constantly undergo self-renewal to give progeny cells that initiate spermatogenesis. These are the only self-renewing population of cells in the adult body capable of contributing to subsequent generations. Stem cells isolated from donor males have restored fertility when transferred into sterile recipients (Brinster and Zimmerman, 1994). Following the development of a method for the reliable transfer of foreign DNA into male germ cells, it will be possible to make transgenic animals using this route.

4.4 Limitations and complications of germline transgenesis

In a transgenic animal, the genetic change represented by the transgene is manifested throughout development, from conception onwards. The lack of temporal or spatial specificity can complicate the interpretation of the phenotypes resulting from a transgenic experiment for three reasons:

1. The overall phenotype observed may be a summation of transgene effects in different tissues at different times. The parts of the overall effect might be very difficult to dissect.
2. An early or severe effect of a transgene might preclude the study of subsequent or downstream processes involving the same gene.
3. Transgenic animals are genetic **'reactionisms'.**

4.4.1 Consequences of reactionism

probs with KO

- A phenotype may be due to endogenous genes being switched on or off in response to the transgene or knockout, rather than being a direct effect of the primary genetic lesion.
- Many investigators have invested effort and resources into the generation of a knockout mouse only to find that the **homozygous** null mutant animal has no overt or obvious phenotype. The knocked-out gene may be 'redundant' – other genes, possibly of the same family, may take over the function of the knocked-out gene.

Other possible reasons for the lack of a phenotype are that:

- The gene may not be completely knocked out. Residual activity might be present because a functional exon was not removed and is expressed through cryptic mechanisms.
- The gene may not have a function. Expression and function are not the same thing.
- Our ignorance of the function of the gene may mean that we do not have a suitable assay for any subtle phenotype resulting from a null mutation.

4.4.2 Inducible and cell-specific transgenesis

Considering these problems, considerable effort is being directed towards the development of transgenic systems that enable lesions to be confined to a particular cell-type or that are temporally controlled. Ideally, a transgenic lesion should be both cell-specific and inducible.

4.5 Recent technological advances

CRE

4.5.1 Site-specific recombinases

The Cre (causes recombination) protein is a site-specific recombinase from bacteriophage P1, a virus that infects *E. coli*. Cre recognises a 34 b site called *loxP* (*lox* = locus of recombination). The recombinase binds to the

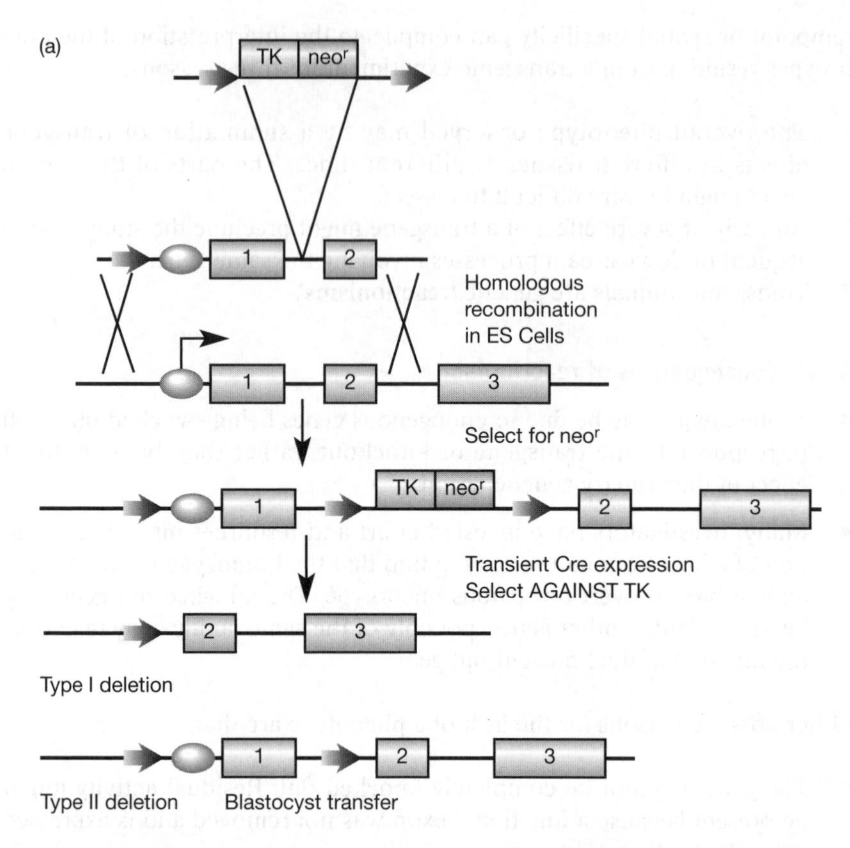

Figure 4.3 'Floxing' a gene. Unlike a conventional gene targeting procedure, the selection markers are eliminated from the mutant locus. This is a two-step process. In the first step, three directly repeated *loxP* sites and two selectable markers – the neomycin resistance gene (neo^r) and HSV-*tk* – are introduced into the flanking regions of the targeted gene by homologous recombination. Following selection for neo^r, ES clones bearing homologous recombination events are identified by genomic analysis. The second step involves the transient expression of Cre in the genetically modified ES cells followed by ganciclovir treatment, which selects against the presence of HSV-*tk*. Three types of recombination product result from Cre activity. A type I deletion results in the deletion of the entire gene from the genome. A type two deletion results in the elimination of the HSV-*tk* gene, leaving the endogenous gene flanked by two directly repeated *loxP* sites (the gene is now 'floxed'). Type III deleted cells, in which the endogenous gene is lost, leaving the HSV-*tk* gene, will die in the presence of ganciclovir. In this example, a type II deletion results in exon 1 being flanked – 'Floxed' – by two *loxP* sites. Type II deleted ES cells are then transferred to the blastocyst resulting, following germline transmission and crossing of heterozygotes, in the generation of mice homozygous for the floxed locus.

13 b inverted repeat elements that flank an 8 b asymmetric core sequence. The core sequence is not involved in recombinase binding, but is the site of crossing over, and the asymmetry of the core gives directionality to the *loxP* site.

As part of the life cycle of the P1 bacteriophage, the genome replicates as a single copy circular plasmid. Cre resolves plasmid dimers into monomers, increasing the number of segregating molecules and preventing

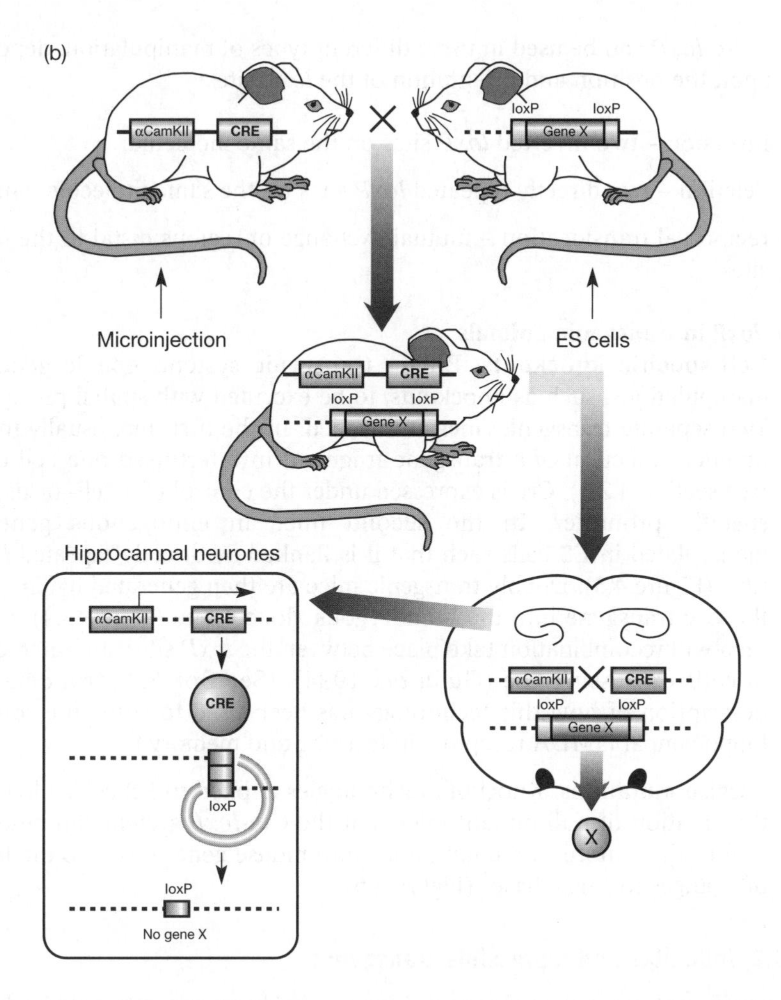

Figure 4.4 Cell-specific knockouts in mice. Two separate transgenic lines are required. The first line, bearing a transgene consisting of a cell-specific promoter driving the expression of Cre, is made by microinjection (see section 4.3.1). In this case, the cell-specific promoter is derived from the αCamKII gene, which is active in hippocampal neurons (Chapter 8). A second line, bearing the 'floxed' gene of interest (Gene X), is derived as described in Figure 4.3. Doubly transgenic mice are then generated by crossing the Cre transgene into the homozygous floxed line. Only then will recombination take place between the *loxP* sites, and then only in cells expressing Cre. In this case, the gene deletion is confined to hippocampal neurones; the rest of the brain continues to express Gene X.

plasmid loss. Cre acts on the *loxP* sites, one of which is present in each genome. Dimers thus have two directly repeated *loxP* sites. Recombination between these sites effectively excises the DNA between them, leading to two circular monomers. Cre-mediated recombination is reversible; therefore the reverse reaction between monomers is possible, but is less favoured. Importantly, supercoiled DNA is not required for Cre activity, and linear molecules are adequate substrates.

Cre-*loxP* can be used in three different types of manipulation, depending upon the position and orientation of the *loxP* sites:

- inversion – two inverted *loxP* sites on the same molecule;
- deletion – two directly repeated *loxP* sites on the same molecule; and
- reciprocal translocation – mutual exchange of regions distal to the *loxP* sites.

Cre-*loxP* in transgenic animals

- **Cell-specific knockouts.** Binary transgenic systems enable genome manipulations, such as knockouts, to be executed with spatial precision. Two separate transgenic lines are needed. In the first line, usually made by microinjection of a transgene fragment into fertilised one-cell eggs (see section 4.2.1), Cre is expressed under the control of a cell- or tissue-specific promoter. In the second line, an endogenous gene is manipulated in ES cells such that it is flanked by directly repeated *loxP* sites (Figure 4.3). Doubly transgenic mice are then generated by crossing the Cre transgene into the homozygous 'floxed' line (Figure 4.4). Only then will recombination take place between the *loxP* sites, and then only in cells expressing Cre (Gu *et al.,* 1994). (See Box 8.2 for a detailed description of how this technology has been used to study the role of hippocampal NMDA receptors in learning and memory.)
- **Precise mutations.** Knockout technologies applied to ES cells allow for the creation of null mutant mice, but the Cre-*loxP* system can be used to introduce more subtle mutations into mouse genes down to the level of changes to single bases (Figure 4.5).

4.5.2 Inducible and repressible transgenes

The temporal control of transgene expression has been attempted with a number of different promoter and inducer systems. The aim of such systems is to:

- isolate transgene expression from complicating developmental effects; and
- enable physiological or behavioural parameters to be examined before, during and after transgene induction on the same groups of animals.

The most effective system described to date is the tetracycline (Tet) gene regulation system. The advantages of this system are that:

- Tet, or analogues thereof, are readily delivered to the animal in drinking water;
- Tet can access all tissues, and can cross the blood–brain barrier; and
- effective doses of Tet have no toxic or gross physiological effects on the animal.

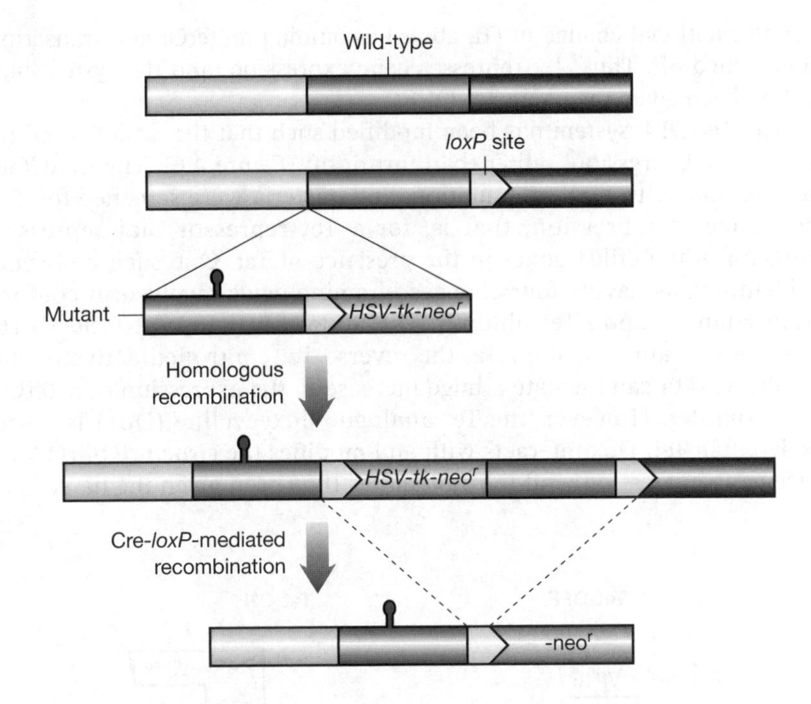

Figure 4.5 Introducing precise mutations into a gene using Cre/*loxP*. The two-step strategy involves the replacement of a gene segment with a homologous segment contained within the targeting vector which bears a mutation of choice. In the first step, conventional homologous recombination is used to replace an endogenous gene with a construct consisting of a mutated gene segment and two selectable markers – HSV-*tk* and the neomycin resistance gene (neo^r). The selectable markers are flanked by directly repeated *loxP* sites. In the targeted locus, the gene segment is replaced with the mutant, downstream of which is the selection cassette, flanked by *loxP* sites. The second step involves the transient expression of Cre in the genetically modified ES cells followed by ganciclovir treatment, which selects against the presence of HSV-*tk*. This results in the deletion of the selectable marker. The targeted gene now bears the mutation of choice. All that remains of the mutagenesis process is a single *loxP* site.

Tetracycline gene regulation

The Tet gene regulation system is based on the Tet resistance (Tet^r) operon from *E. coli,* tn10. Tet resistance in Gram-negative bacteria is mediated by proteins effecting the active efflux of the drug from the cell. The expression of the efflux genes is regulated by the Tet repressor protein, which binds to two operator (TetO) sequences in the absence of tetracycline. High-level expression of the efflux genesis is induced by Tet, which binds to the repressor and eliminates operator recognition.

This system has been adapted to mammalian use by generating the Tet-controlled trans-activator protein (tTa), consisting of a fusion between the Tet repressor and the activator domain of the powerful herpes simplex transcription factor VP16 (Figure 4.6). In the absence of Tet, tTA binds to, and strongly transactivates, artificial genes controlled by TetO sequences linked to a basal, or minimal, eukaryotic promoter. In the presence of Tet, the conformational change in tTa abolishes binding to TetO, and transcription is turned off. Thus, Tet represses gene expression, and the system has therefore been given the name Tet-OFF.

The Tet-OFF system has been modified such that the drug is used to turn on gene expression, rather than turn it off (Figure 4.6). The tn10 Tet^r gene was subject to random mutation, and bacteria were screened for Tet dependence of repression, that is, for a Tet repressor that represses expression of the efflux genes in the presence of Tet. One such molecule was identified as having four changes in amino acids involved in conformation changes upon Tet binding. The mutant was fused to the VP16 activation domain to make rtTa, the reverse Tet-controlled activator. In cell culture, rtTa can mediate a huge increase in the expression of a TetO-basal promoter. However, the Tet analogue doxycycline (Dox) is more effective than Tet. Dox interacts with and modifies the structure of rtTA to induce gene expression, and the system has thus been given the name Tet-

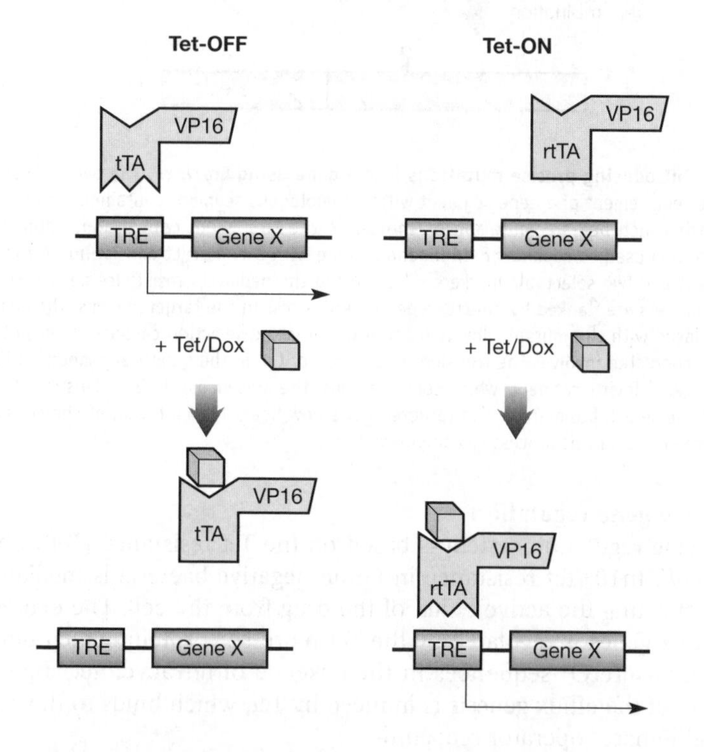

Figure 4.6 Tet-OFF and Tet-ON. Tet-OFF: The Tet-controlled *trans*-activator protein (tTa) consists of a fusion between the Tet repressor and the activator domain of the powerful herpes simplex transcription factor VP16. In the absence of Tet, tTA binds to, and strongly transactivates, artificial genes controlled by TetO sequences linked to a basal, or minimal, eukaryotic promoter (TRE – Tet response element). In the presence of Tet, the conformational change in tTa abolishes binding to TetO, and transcription is turned off. Tet-ON: the reverse Tet-controlled *trans*-activator protein (rtTa) is unable to bind the TRE in the absence of the Tet analagoue doxycycline (Dox). In the presence of Dox, rtTA binds to, and strongly transactivates, artificial genes controlled by the TRE.

ON. More information about Tet-ON and Tet-OFF can be found at the web site of Clontech Laboratories Inc. **(http://www.clontech.com/clontech/Catalog/GeneExpression/Tet.html)** and at Dr Hermann Bujard's web site **(http://www.zmbh.uni-heidelberg.de/bujard/homepage.html)**.

Tetracycline-regulated gene expression in transgenic animals

The Tet gene control system works effectively in transgenic animals (see Box 8.4). Two transgenic lines are required. In the first line, either tTa or rtTa are expressed under the control of a mammalian promoter – probably a cell-specific promoter. In the second line, a coding sequence of choice is expressed under the control of a basal eukaryotic promoter controlled by TetO. To facilitate Tet or Dox gene regulation, the two lines are mated to generate doubly transgenic mice. In Tet-OFF mice, expression of the coding sequence is repressed to negligible levels by Tet or Dox. In Tet-ON mice, expression of the coding sequence is activated many hundreds of fold by Dox. Regulation is confined to particular cells, being governed by the specificity of the promoter used to direct the expression of tTa or rtTa (Kistner *et al.*, 1996).

4.5.3 Inducible knockouts

The inducible knockout of a gene has been demonstrated in transgenic mice (Kuhn *et al.*, 1995). The MX1 promoter was used to express Cre in an inducible fashion in transgenic mice. MX1 is induced by interferon-α or interferon-β, or by the IFN inducer polyinosinic acid-polycytidylic acid (pI-pC). In a second line of mice, the gene encoding DNA polymerase β gene was floxed. Bi-transgenic mice were then given pI-pC, which induced Cre expression resulting in the deletion of the DNA polymerase β gene.

4.5.4 Inducible and cell-specific knockouts

Ideally, a knockout should be both cell-specific and inducible, and the technology now exists to enable this. A cell-specific, inducible knockout requires the generation of mice bearing three genetic modifications:

1. A cell-specific promoter directing the expression of rtTA.
2. A TetO-controlled basal mammalian promoter directing the expression of Cre.
3. A floxed gene.

Each of these components on their own has no effect on the mouse. Indeed, in combination, no effect will be apparent until Dox is fed to the animal. Dox will induce a conformational change in the rtTa protein, facilitating its interaction with TetO, and the induction of the minimal mammalian promoter. The TetO-minimal promoter will direct the expression of Cre, which will mediate deletion of the floxed gene. Cells that do not express rtTa will remain unaltered.

4.6 Alternatives to gene knockouts

4.6.1 Expression of antisense RNA

Watson–Crick base-pairing of an antisense RNA corresponding to a mRNA can potentially inhibit gene expression. The antisense RNA could interfere at one of a number of levels in the overall process of gene expression:

- transcription – the generation of the primary transcript;
- processing of the primary transcript;
- transport of the RNA from nucleus to cytoplasm; and
- translation

Alternatively, antisense RNA could promote mRNA degradation by forming a substrate for nucleases specific for double-stranded RNA.

Whatever the potential mechanism, there have not been many reported successes with the approach. One group has sought to block activity of glucocorticoid receptor by expression of an antisense glucocorticoid receptor RNA (Pepin *et al.*, 1992). Although expressed under the control of the neurofilament l gene promoter, antisense RNA was present throughout the body, resulting in a 50–70% decrease in glucocorticoid receptor mRNA, and a significant reduction in glucocorticoid binding in a number of brain regions. The main phenotype observed was gross obesity.

4.6.2 Dominant-acting agonists or antagonists

Studies on intracellular signalling pathways have resulted in the development of dominant-acting mutants of signalling molecules that are antagonists or agonists of their wild-type counterparts. These mutants can be expressed under the control of cell-specific regulatory sequences, resulting in the inhibition or the constitutive activation of particular pathways (see Chapter 8 for a description of how such molecules have been used to study **learning** and **memory**).

4.6.3 Cell ablation

Cell-specific promoters can be used to kill specific cell types through the expression of highly toxic proteins such as the A chain of diphtheria toxin (DTA). Diphtheria toxin is made up of of two chains, A and B. The B chain (DTB) mediates the entry of the toxin into the cell, while the A chain kills the cell by inactivating elongation factor 2, leading to inhibition of translocation and GTPase activity, and an inhibition of protein synthesis. DTA is highly toxic and one molecule can kill a cell. In transgenic experiments, only the A chain is expressed. Without the B chain, the DTA can only kill the cell that expresses it. DTA released from dead cells is not harmful to neighbouring cells because it cannot enter them without DTB. DTA has been expressed in mice under the control of 310b of rat growth hormone promoter, which mediates specific expression to growth hormone-producing cells in the anterior pituitary (Behgringer *et al.*,1988). These transgenic mice have no growth hormone-producing cells, no detectable circulating growth hormone and, as a result, suffer from dwarfism.

Inducible cell ablation can be achieved through the cell-specific expression of the thymidine kinase gene from herpes simplex virus (HSV-TK). Cells expressing HSV-TK acquire sensitivity to drugs such as the nucleoside analogue, ganciclovir. Ganciclovir is harmless to animals and animal cells, but metabolites of ganciclovir and HSV-TK can kill dividing cells. Thus, if ganciclovir is given to an animal, cells expressing HSV-TK will die, but cells that do not express HSV-TK will survive. HSV-TK phosphorylates nucleoside analogues to give the nucleoside monophosphate. This is phosphorylated by cellular kinases to the nucleoside triphosphate, which is incorporated into DNA, and inhibits further cellular DNA synthesis by blocking elongation. The growth hormone promoter has been used to direct the expression of HSV-TK to growth hormone-producing cells of the anterior pituitary gland (Borrelli *et al.*, 1989). Mice treated with ganciclovir through post-natal development develop as dwarfs, but this is reversible. If ganciclovir is withdrawn, stem cells repopulate the pituitary with mature growth hormone-producing cells.

4.6.4 Recombinant antibodies

The specificity of antibody binding to an epitope can be used to interfere with the functioning of a protein. Genes encoding antibodies can be expressed in specific cell types using appropriate regulatory sequences resulting in binding to, and inhibition of, the target protein. Substance P has been thus targeted. Substance P is a neuropeptide with a number of putative neurotransmitter and neuromodulator functions in the CNS. The regulatory sequences of the neuronal VGF gene were used to direct the expression to the mouse CNS of an antibody that recognises substance P. The transgenic mice exhibited motor deficits and a marked inhibition of neurogenic inflammation (Piccioli *et al.*, 1995).

4.7 Genetic background

Complex traits involve interactions between large numbers of different genes, and each gene can be represented in a population by different alleles that can modify the expression of the trait. It has been found that knockout phenotypes can vary dramatically depending on the genetic background. A good example of this is the epidermal growth factor receptor (EGFR) gene knockout. EGFR is a member of the tyrosine kinase family of receptors that is widely expressed during development and in the adult. The EGFR gene knockout has complex effects in different mouse strains (Miettinen *et al.*,

1995; Sibilia and Wagner, 1995; Threadgill *et al.*, 1995) Most knockouts are made in ES cells derived from the 129/SV inbred strain of mice and, in this strain, the EGFR knockout homozygotes die in mid-gestation due to placental defects. However, on the 129/SV × C57Bl/6 background, the mice survive until birth, and the newborns have open eyes, rudimentary whiskers, immature lungs and epidermal defects – all of which correlate with EGFR expression. On the CD1 background, the knockouts live for over three weeks after birth and show defects in skin, liver, brain, kidney and gut. However, in the CF-1 background, the knockout homozygotes die before blastocyst implantation due to degeneration of the inner cell mass. It is thus evident that EGFR pathways are regulated differently in different mouse strains. Perhaps different tyrosine kinase receptors interact with, or compensate for, EGFR, but these effects are strain-dependent.

4.8 Summary

- The mammalian genome can be manipulated at will:
  - new genes can be added;
  - endogenous genes can be removed (knockouts);
  - endogenous genes can be subtly mutated down to the level of a single base;
  - designer genetic lesions can be rendered cell-specific, inducible, or both.
- Gene knockout experiments have revealed the phenomenon of redundancy – other genes, possibly of the same family, may take over the function of the knocked-out gene.
- Genetic background can have a profound effect on the phenotype of a transgenic mouse.

Further reading

HOGAN, B., BEDDINGTON, B., CONSTANTINI, F. and LACY, E. (1994) *Manipulating the Mouse Embryo: a Laboratory Manual*. Cold Spring Harbor Laboratories, Cold Spring Harbor, New York.

The 1986 publication of Manipulating the Mouse Embryo catalysed the interaction between molecular biology and mammalian embryology. The second edition of this classic manual contains new sections on the production and analysis of transgenic mice, the manipulation of preimplantation embryos to generate chimeras, the culture and manipulation of embryonic stem cells, including gene 'knockouts,' and techniques for visualising genes, gene products and specific cell types.

MURPHY, D. and CARTER, D.A. (eds) (1993) *Methods in Molecular Biology*, Vol 18. *Transgenesis Techniques: Principles and Protocols*. Humana Press, Totowa, NJ.

More of a techniques-based volume that the classic Hogan text. Includes reviews of application of transgenesis and detailed descriptions of systems not covered by Hogan (fish, rat, sheep).

GLOVER, D.M. and HAMES, D. (eds) (1995) *DNA Cloning 4 – A Practical Approach. Mammalian Systems*. IRL Press at Oxford University Press, Oxford.

An up-to-date and clear methods book which covers rodent transgenesis (rats and mice; microinjection and embryonal stem cells) as well as viral gene transfer (retroviruses, adenoviruses, herpesviruses).

CLOUTHIER, D.E., AVARBOCK, M.R., MAIKA, S.D., HAMMER, R.E. and BRINSTER, R.L. (1996) Rat spermatogenesis in mouse testis. *Nature*, **381**: 418–421.

Whatever next! Immunodeficient mouse hosts undergo rat spermatogenesis after receiving transplants of rat spermatogonia.

CHAPTER FIVE

Somatic transgenesis

Key points

- Antisense
- Viral-mediated gene transfer
    - Retroviruses
    - Adeno-associated virus
    - Herpes simplex virus
    - Adenovirus
- Liposomes
- Cationic polymers
- Naked DNA
- Biolistics
- Gene therapy
    - Parkinson's disease

5.1 Introduction

Somatic gene transfer refers to the introduction of new genes into the somatic cells of an organism. The germline is not affected and the genetic change will not be passed on to subsequent generations. Somatic gene transfer has matured into a valuable tool in basic molecular neuroscience research, and has attracted considerable interest in relation to its possible clinical applications.

5.2 Antisense oligonucleotides

Attempts have been made to block gene expression by the application of large quantities of antisense single-stranded DNA oligonucleotides corre-

sponding to the sequence of a target gene or its mRNA
(http://www.hybridon.com/graphic_version/antisense/block.html
As such, antisense has the potential to be an alternative to germline knock-out technologies.

Despite being widely used, antisense is highly controversial and remains unproven in most circumstances. It is likely that antisense molecules are able, in some circumstances, to interfere specifically with their intended targets. Such an outcome demands that the target RNA is accessible to the antisense molecule, but such cases are rare, as mRNAs are tightly packaged by proteins and secondary structure into ribonucleoproteins in the cell. However, it has never been proven that an antisense molecule can knock out just one gene product, and that all of the other expressed genes in the target cell remain unaltered. Further, antisense molecules, particularly oligonucleotides which have been chemically modified to increase their stability and efficacy, can have profound non-sequence-specific effects on cells. Modified oligonucleotides can bind avidly to proteins, and breakdown products can inhibit cell proliferation. A biological effect seen as a result of applying an antisense molecule might therefore be the result of:

- a specific Watson and Crick interaction with its intended target;
- a specific Watson and Crick interaction with an unintended, but unidentified, target;
- a non-antisense interaction with another RNA or RNAs;
- an interaction with protein; or
- a non-specific effect on cell proliferation or metabolism.

A 'Cutting Edge Debate' on the validity, or otherwise, of the antisense approach can be found on **http://biomednet.com.**

5.3 Viral-mediated gene transfer

Viruses have evolved to deliver foreign nucleic acid efficiently into a host cell. As such modified viruses make ideal vectors for the delivery of foreign genes into cells. A number of different classes of virus have been used as gene delivery systems, including:

- Retroviruses (see Chapter 4;
  http://www.micro.msb.le.ac.uk/335/Retroviruses.html)
- Adeno-associated virus (AAV;
  http://www.micro.msb.le.ac.uk/335/Parvoviruses.html)
- Herpes simplex virus (HSV;
  http://www.micro.msb.le.ac.uk/335/Herpesviruses.html;
  http://fiona.umsmed.edu/~yar/hsv.html)

- Adenovirus (Ad; **http://fiona.umsmed.edu/~yar/adeno.html; http://www-micro.msb.le.ac.uk/335/Adenoviruses.html; http://www.uct.ac.za/depts/mmi/stannard/adeno.html**)

5.3.1 Retroviruses

Gene transfer vectors derived from retroviruses are described in section 4.3.2. Effective gene transfer systems for the CNS have recently been derived from the Human Immunodeficiency Virus (HIV), the Lentivirus **(http://pavlakislab.ncifcrf.gov/HIVreview17.html)** which causes AIDS.

5.3.2 Adeno-associated virus (AAV)

Parvoviruses are among the smallest, simplest eukaryotic viruses. Defective adeno-associated viruses (dependoviruses), such as AAV, are reliant on Ad or Herpesvirus helper functions to replicate. AAV consists of a linear, non-segmented, single-stranded DNA genome of around 5 kb (Figure. 5.1a), and three capsid proteins, VP1-3. The genome is flanked by palindromic sequences (inverted terminal repearts, or ITRs) which form hairpins. The ITRs are important in the site-specific integration of AAV DNA into chromosome 19, and are essential for the initiation of genome replication. The right part of the genome (cap) codes for the structural capsid proteins, while the left part of the genome (rep) codes for the non-structural proteins that are required for parvoviral DNA replication. **Replication** occurs in the nucleus and is highly dependent on cellular functions such as the host cell DNA polymerase (Figure. 5.1b). AAV expression is poorly understood, but requires the helper function provided by the early (regulatory) region of adenovirus. This function appears to modify the cellular environment rather than interacting directly with the AAV genome. In the absence of a helper virus, AAV enters a latent state in which the virus genome is integrated into the host cell DNA. Latency can be rescued by a subsequent adenoviral infection.

AAV vectors are prepared by replacing AAV genes with the gene of interest, flanked by ITRs (Figure 5.1a). Defective AAV vectors (either cap-deleted [type 1] or rep- and cap-deleted [type 2]) are used to generate infectious AAV particles. Vectors are transfected into packaging cells that express AAV cap or AAV cap and rep (and thus complement the missing genes in these defective vectors). The packaging cells are infected with helper adenovirus. The adenovirus allows AAV to replicate, and the AAV particles are then purified away from the adenovirus particles by heating them (AAV is heat stable, whereas adenovirus is not) and then separating intact viral particles from cellular debris using density gradient centrifugation.

Advantages

- Site-specific integration.
- Stable latent state.

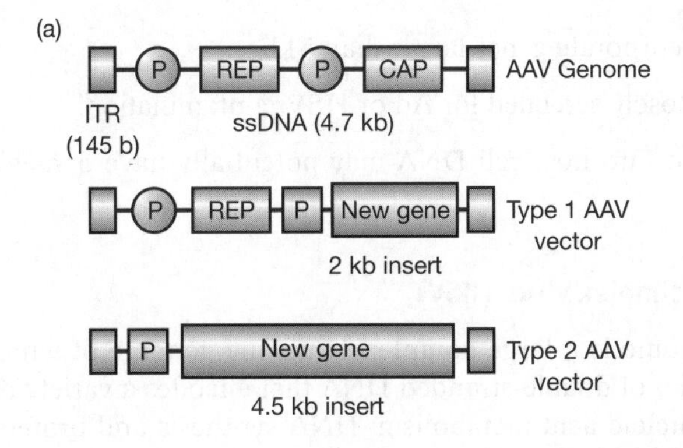

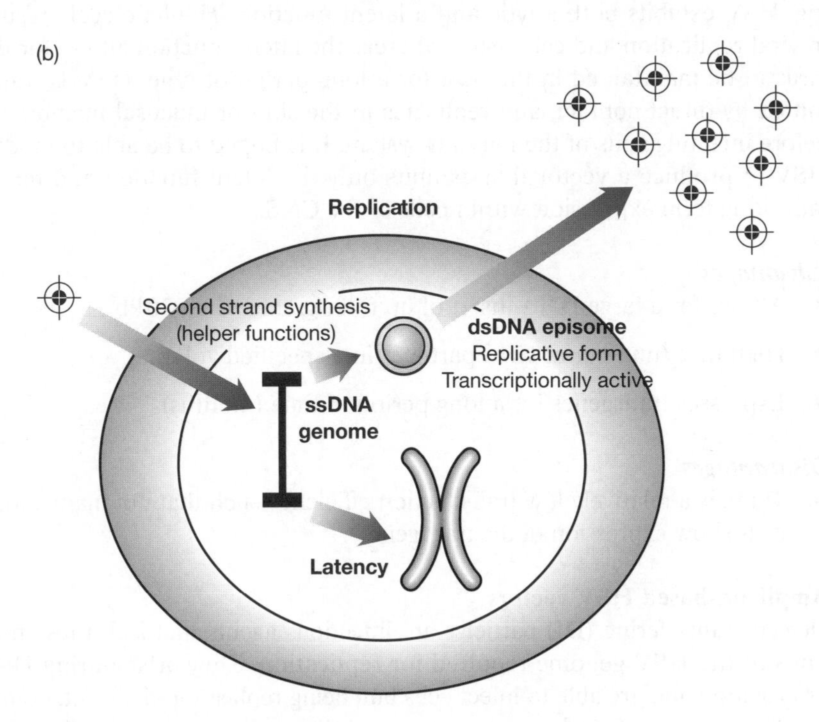

Figure 5.1 (a) Structure of the AAV genome and of AAV gene therapy vectors. (b) AAV relication and latency.

- Lack of an immune response – no viral antigens are present.
- Not associated with human disease.
- Infect a variety of dividing and non-dividing cells.
- Because of the simplicity of the genome, vectors can be constructed which will not express any (undesirable) viral gene products.

Disadvantages

- Cannot incorporate genes larger than 5 kb.
- Must be closely screened for Ad or HSV contamination.
- Integration into host cell DNA may potentially have damaging consequences.

5.3.3 Herpessimplex virus (HSV)

The HSV genome is a large complex virus, the genome of which is composed of 160 kb of double-stranded DNA that encodes a variety of enzymes involved in nucleic acid metabolism, DNA synthesis and protein processing. HSV exhibits both a lytic and a latent function. The lytic cycle results in viral replication and cell death, whereas the latent function allows for the virus to be maintained in the host for a long period of time. HSV is transmitted by direct contact, and replicates in the skin or mucosal membranes before infecting cells of the nervous system. It is hoped to be able to modify HSV to produce a vector that exhibits only the latent function and maintains long term expression within cells of the CNS.

Advantages

- Allows for a large DNA insert of up to or greater than 20 kb.
- High titre (number of virus particles in a specified volume).
- Expresses transgenes for a long period of time (months).

Disavantages

- There is a relatively low transduction efficiency such that not many target cells show expression of the transgene.

Amplicon-based HSV vectors

Defective interfering (DI) particles are infectious agents that lack those portions of the HSV genome required for replication. They arise during HSV propagation and are able to infect cells but, being replication-deficient, do not propagate further. Based on this observation, it was shown that plasmid DNA containing packaging signals could also be incorporated into DI particles.

Amplicon or DI vectors consist of plasmid DNA which contains the HSV packaging signals and a cloned gene of interest. The vector is transfected into any cell line which will allow for efficient HSV propagation. The cells are then infected with helper HSV virus. The DI particle containing the recombinant vector is generated along with helper virus during the viral propagation process.

Advantages

- Easy to prepare.

Disadvantages

- DI particle stocks are contaminated with helper virus. Packaging cell lines are being developed to enable virus-free DI particle maturation.

5.3.4 Adenoviruses

Adenoviruses (Ads) were discovered in 1953 as investigators attempted to identify the causative agents of the common cold. Of the over 100 serotypes known, human Ad5 has been the most extensively studied. The Ad genome consists of linear, non-segmented, double-stranded DNA, 30–38 kbp that encodes 30–40 genes (Figure 5.2a).

Adenoviral infection is a highly complex process. Infection is initiated by the virus binding to the cellular receptor. Internalisation occurs via receptor-mediated endocytosis followed by release from the endosome. The viral capsid undergoes disassembly as it journeys to the nucleus. The viral DNA does not integrate into the host genome but remains in an episomal state. Viral DNA replication and transcription are complex, and viral replication occurs only in the nucleus of infected cells. The replication cycle has two phases: the early phase – the events before viral replication; and the late phase – the events after the initiation of DNA replication. During the early phase, the genes of four non-contiguous early regions of the genome are expressed. The E1a region is the first gene to be transcribed upon nuclear entry and is essential for viral replication. The E3 region has a key role in pathogenesis. After onset of DNA replication, the major late promoter drives the expression of the late proteins through a complex pattern of splicing and processing. The late-encoded virion proteins assemble around the replicated viral DNA until the lysis of the host cell.

Replication-deficient Ad vectors

To make an Ad capable of delivering new genes to a host cell, two important technical advances have been necessary:

1. The development of Ad5 derivatives that are able to accommodate insertions or substitutions. It has been necessary to make deleted forms of the viral genome because the virion can only package a maximum amount of DNA corresponding to 105% of the wild-type genome size.
2. The development of Ad5 derivatives that are replication-incompetent. That is, they are able to infect cells and deliver their DNA, but within the host that DNA will not replicate.

These ends were achieved by deletion of the E1 and E3 regions (Figure 5.2a). This freed genome capacity for the insertion of transgenes and, because the E1 region is necessary for viral DNA replication, the resulting virus cannot be propagated autonomously. The E3 region appears not to be required for viral growth in culture, and removing it allows for a larger DNA insert to be incorporated into the vector. Replication-deficient viruses

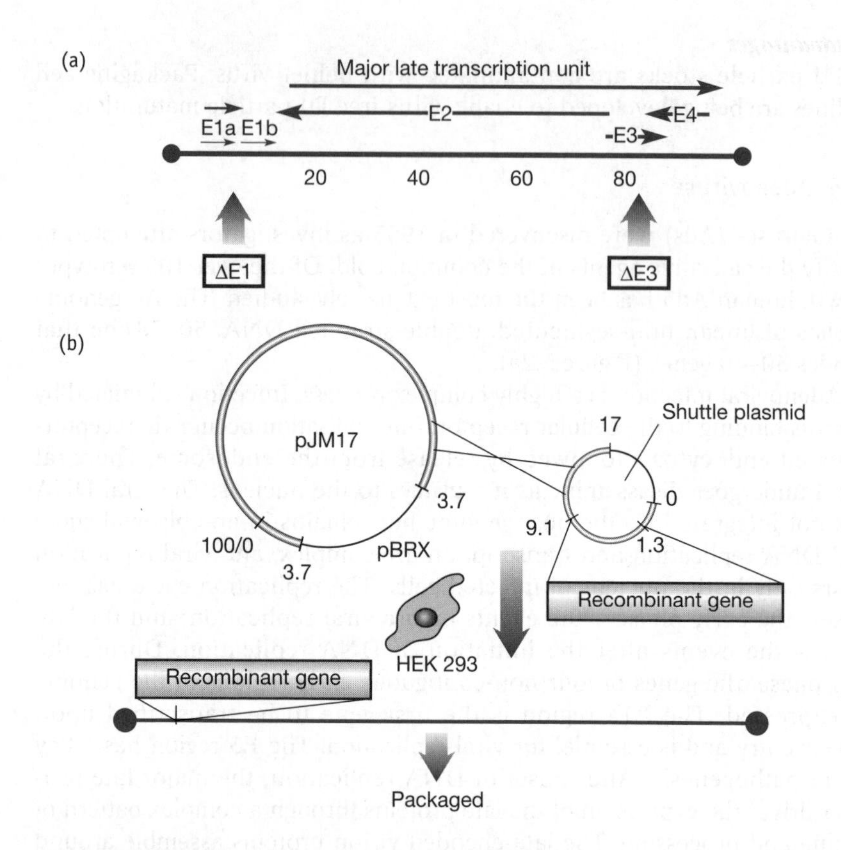

Figure 5.2 (a) Structure of the Ad genome. Replication deficient Ad vectors were prepared by deletion of the E1 and E3 regions.
(b) Preparation of a recombinant Ad. 1. The recombinant gene is cloned into a shuttle plasmid which contains a subfragment of the viral genome that spans the E1 deletion. 2. The subfragment, now bearing the transgene, is reconstituted into an infectious virus by recombination with another plasmid bearing the rest of the Ad genome (e.g. pJM17). pJM17 is also devoid of the E1 region, but does have the rest of the viral coding sequence needed to produce a virion. 3. The recombinant shuttle plasmid and pJM17 are co-transfected into HEK 293 cells, which provide E1 *in trans*. Recombination between the two plasmids results in the generation of an intact viral genome which is packaged into infectious particles.

are propagated in a cell line, HEK 293 – a human embryonic kidney fibroblast line that has been transformed by Ad5, and thus provides the E1 gene products *in trans*.

Advantages

- The Ad viral particle is stable.
- The Ad viral genome is stable, does not undergo rearrangement, and the inserted foreign genes are maintained without change through many rounds of replication.
- The viral genome is easy to manipulate using standard recombinant DNA techniques.

- The virus replicates efficiently in permissive cells to high titres – each infected cell produces 1000–10 000 pfu (plaque-forming units, or infectious virion particles).
- Ads can infect many types of cell.
- Ads can infect non-replicating cells, including post-mitotic neurones.
- Ads do not integrate into the host genome and cannot exert any insertional mutagenic effects.
- Ads can accommodate transgenes of up to 8 kb.

Disadvantages

- Ad expression is transient since the viral DNA does not integrate into the host genome.
- Viral proteins are expressed following Ad vector administration into the host.
- Viral protein expression can elicit a host immune response which eliminates the infection and limits the duration of transgene expression.

The immune response to Ad

The main problem encountered with Ad gene transfer is the immune response mounted by the host against the virus. However, it turns out that this is not a great problem in the brain, which is protected by the blood–brain barrier. Indeed, it has now been demonstrated that Ad vectors can mediate the long-term correction of inherited genetic defects affecting the CNS. Injection of an Ad carrying a vasopressin cDNA has corrected the symptoms of diabetes insipidus in the vasopressin-deficient Brattleboro rat (Geddes *et al.*, 1997) for up to four months.

5.4 Liposomes

Cationic liposomes consist of a positively charged lipid and a co-lipid that interact with the negatively charged DNA molecules to form a stable complex. Complexes are taken up by neurones, probably by endocytosis. The efficiency of liposomal delivery of DNA can be increased by the incorporation of viral proteins to form virosomes. The viral proteins interact with cellular receptors, facilitating endocytosis. For general information about liposomes, see **http://www.unizh.ch/onkwww/lipos.htm.**

Advantages

- Liposomes can complex both with negatively and positively charged molecules.
- Liposomes offer a degree of protection to the DNA from degradative processes.

- Liposomes can carry large pieces of DNA.
- Liposomes can be targeted to specific cells or tissues.
- No immune response.
- Free of viral contamination.

Disadvantages
- Low transfection efficiencies.
- Transient gene expression.
- Some cellular toxicity.

5.5 Cationic polymers

Cationic polymers, such as polyethylenimine (PEI) show great promise as gene transfer vectors, particularly in the CNS (Abdallah *et al.*, 1996). Every third atom of PEI is a protonatable amino nitrogen atom which makes the polymeric network an effective 'proton sponge' at virtually any pH. The efficiency of gene transfer is thought to rely on extensive endosome and lysosome buffering that protects DNA from nuclease degradation, and consequent lysosomal swelling and rupture that provides an escape mechanism for the PEI/DNA particles.

Advantages
- Cationic polymers efficiently complex with negatively charged molecules such as DNA.
- Cationic polymers offer a degree of protection to the DNA from degradative processes.
- Cationic polymers can carry large pieces of DNA.
- Cationic polymers can be targeted to specific cells or tissues.
- No immune response.
- Free of viral contamination.
- High transfection efficiencies.
- Long-lasting gene expression.
- Low cytotoxicity.

5.6 Naked DNA

Transgene expression has been observed in the rodent brain following intracerebral injection of naked circular plasmid DNA, or stereotaxic injection of DNA into specific brain regions (Hannas-Djebbara *et al.*, 1997).

Expression was detectable 48 hours after injection, and persisted for up to two months. Interestingly, reporter expression was observed predominantly in neurones when the neurone-specific enolase gene promoter was used, and glial expression was predominant with the glial fibrillary acidic protein gene promoter. The mechanism of plasmid DNA uptake is unknown.

Advantages

- Simple.
- Big DNA fragments can be introduced.

Disadvantages

- Low transfection efficiencies.
- Transient expression.

5.7 Biolistics

Genes can be literally shot into cells. DNA is coated onto 1–3 ^{mu}m-sized gold or tungsten particles, which are placed onto a carrier sheet which is inserted above a discharge chamber (a 'gene gun' – see **http://www.bio-rad.com/972508.html).** At discharge, the carrier sheet accelerates toward a retaining screen which halts the carrier sheet, but allows the particles to continue toward their target. Unfortunately, genes delivered by this method are expressed transiently and considerable cell damage occurs at the centre of the discharge site. *In vivo* applications have predominantly focused on the liver, skin, muscle or other organs which can easily be exposed surgically. However, brain slices in culture can be targeted using biolistics.

5.8 Gene therapy and the CNS

Gene therapy is the new frontier in medicine. The methods described in this chapter are used to deliver new genes to the somatic cells of an intact organism. Neuroscience research will benefit greatly from their application. However¢, increasingly, somatic gene transfer will leave the laboratory and enter the clinic and the hospital in the guise of 'gene therapy'. Gene therapy embodies the hope that diseased somatic tissues can be repaired by intervention with nucleic acid drugs. Vanderbilt University hosts an excellent web site devoted to gene therapy issues **(http://www.mc.vanderbilt.edu/gcrc/gene/index.html).** The latest developments in gene therapy – technical, medical and ethical – can be followed by regular viewing of *'Gene Therapy Weekly'* **(http://www.homepage.holowww.com/x1g.htm).** For some, the prospect of gene therapy raises difficult ethical or moral issues, and the web is a forum for all shades of opinion **(http://dspace.dial.pipex.com/srtscot/srtpage3.shtml; http://www.med.upenn.edu/~bioethic/genetics/articles/1.caplan.gene.therapy.html; http://www.med.upenn.edu/~bioethic/genetics/articles/12.gen.disease.html).**

5.8.1 Gene therapy – factors for consideration

- Is the target organ readily accessible?
- Is the disorder detectable early enough for useful intervention?
- Will benefit be therapeutic or symptomatic?
- What is known about the disorder? How complete is our understanding of the central pathophysiology?
- Is the gene small enough to be accommodated in a therapy vector?
- What is the most appropriate vector system?
- Is the disease fatal?

5.8.2 A case study – Parkinson's disease

Parkinson's disease (PD) is a common, chronic and ultimately debilitating neurodegenerative disorder of the CNS. Information about PD is abundant on the World Wide Web. The following sites provide reliable and accessible general information about PD, and contain links to more pages with more technical content:

- **http://dna2z.com/projects/PD**
- **http://neuro-chief-e.mgh.harvard.edu/parkinsonsweb/Main/IntroPD/NIHPub139.html**
- **http://neuro-chief-e.mgh.harvard.edu/parkinsonsweb/Main/IntroPD/PDintroMenu.html**
- **http://neurosurgery.mgh.harvard.edu/LnkFNCTN.htm#OtherPDInfo**
- **http://www.parkinson.org/**

What is wrong in Parkinson's patients?
Affected individuals suffer from debilitating deficiencies in motor functions that manifest in:

- rhythmic tremors at rest;
- inability to initiate (akinesia) or complete (bradykinesia) routine movements;
- muscle rigidity that leads to jerky movement;
- postural instability;
- lack of facial expression.

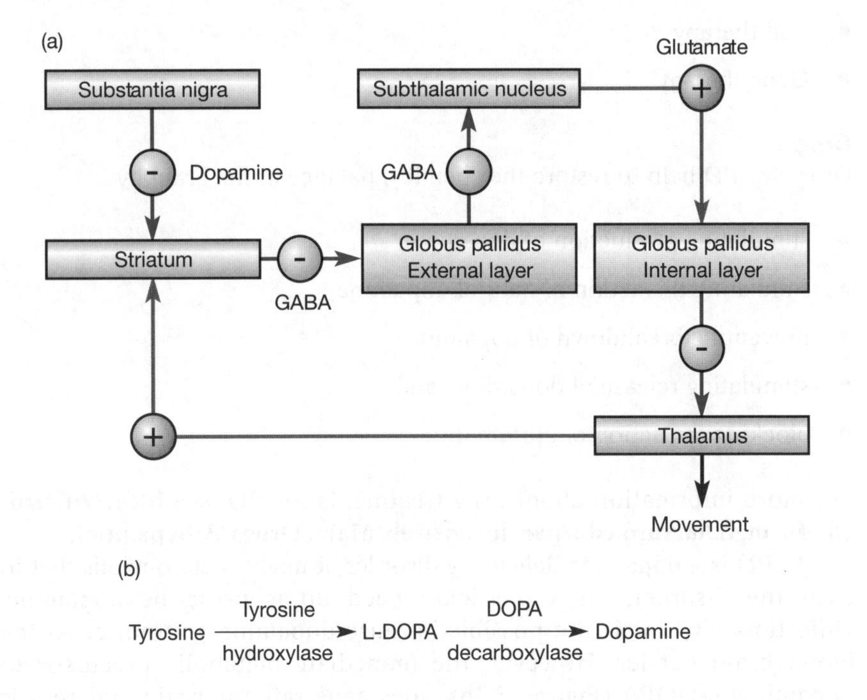

Figure 5.3 (a) The CNS inhibitory (−) and stimulatory (+) pathways involved in the control of normal movement. (b) The dopamine biosynthetic pathway.

The clinical symptoms are caused by a selective loss of pigmented dopamine-producing neurones from the substantia nigra (SN) in the midbrain and a consequent decrease in dopamine at the innervation targets of these neurones, including the striatum. Loss of >80% of these cells results in the classic symptoms of PD. The cause of cell death in PD is unknown. Viral infections, environmental toxins and oxidative stress induced by dopamine metabolites, are all suspected. In contrast, the consequence of dopaminergic neuronal loss for the neural circuits that control movement is well understood. The SN dopaminergic neurones are an integral part of the basal ganglia, a group of forebrain nuclei that play an important role in motor control (Figure 5.3a). In their absence, excessive inhibitory stimuli are sent from the basal ganglia through the globus pallidus to the thalamus. In addition, the loss of a feedback loop between the nigral dopaminergic neurones and the thalamus leads to the release of spontaneous periodic impulses in the thalamus, which are responsible in part for the characteristic tremors.

PD therapy

These include:

- Drugs
- Surgery

- Cell therapy
- Gene therapy

Drugs
Drugs for PD help to restore the chemical balance in the brain by:

- stimulating production of dopamine;
- simulating the action of natural dopamine;
- preventing breakdown of dopamine;
- stimulating release of dopamine; and
- blocking the action of **glutamate.**

For more information about drug treatments for PD, see **http://neurochief-e.mgh.harvard.edu/parkinsonsweb/Main/Drugs/Athena.html.**

As PD is a dopamine deficiency disorder, it might seem obvious that to treat the disorder, all a physician need do is prescribe dopamine. Unfortunately, this is not possible because dopamine will not cross the blood–brain barrier. However, the immediate metabolic precursor to dopamine, L-DOPA (Figure. 5.3b), does penetrate the brain, where it is decarboxylated to dopamine. L-DOPA replacement therapy (oral Levodopa in combination with an amino acid decarboxylase inhibitor) revolutionised the treatment of PD in the early 1970s. However, most of the patients develop motor fluctuations as the disease progresses. As the cells in the substantia nigra continue to die, it becomes increasingly difficult to provide the correct dose of L-DOPA to make up for the missing dopamine. Drug therapy is also complicated by adverse effects; the administration of DOPA leads to non-physiological and intermittent stimulation of striatal neurones bearing dopamine receptors. High doses of L-DOPA can result in symptoms more unpleasant than those of PD. Attempts to treat PD symptomatically have centred around modulation of the resultant increase in glutamatergic output from the subthalamic nucleus (Figure. 5.3a). However, compounds that block glutamate-associated ion channels, such as MK-801 and phencyclidine (PCP, known on the street as 'Angel Dust'), provoke intolerable disassociative anaesthetic effects.

Surgery
A surgical lesion within the internal portion of the globus pallidus (pallidotomy; **http://neurosurgery.mgh.harvard.edu/Pd-pract.htm; http://mcns10.med.nyu.edu/CMD/pallidotomy.html**) can provide symptomatic benefit in PD. Pallidotomy probably exerts its therapeutic effect by reducing the response of the globus pallidus to the excessive glutamatergic stimulation of the subthalamic nucleus, hence reducing the inhibitory input to the thalamus (Figure 5.3a). Pallidotomy is a difficult lesion, as the internal por-

tion of the globus pallidus is flanked by the optic tract medially and the posterior limb of the internal capsule (carrying motor fibres) laterally. The risk of stroke or death from this procedure is estimated to be approximately 5%. However, a successful procedure may significantly decrease the amount of drug-induced dyskinesia, enabling treatment with increased L-DOPA. A case study describing the benefit one patient received from a pallidotomy operation is presented at **http://neurosurgery.mgh.harvard.edu/Pdpallid.htm.** The site **http://pallidotomy.com/** has some useful general information about PD, and some Quicktime video clips illustrating the benefits of pallidotomy.

Cell therapy

The rationale for the new emerging transplantation treatment for PD is the replacement of dopamine neurones that have died. Various sources of cells are being investigated:

- **Allogeneic (within species) – fetal brain**. Human clinical trials involving the transplantation of dopaminergic neurones from aborted human fetuses were conducted in the late 1980s. Promising results were observed, including, in some cases, marked symptomatic recovery of PD patients paralleled by restitution of dopamine synapses in the host brain and reduced need for, or more effective action of, L-DOPA. Other patients, however, showed no benefits and continued to decline. Problems with this approach are:
  - inconsistent responses;
  - ethical and practical concerns of using fetal tissue;
  - difficulty in obtaining a sufficiently large number of human fetal neurones – up to 10 human fetuses are needed per graft; and
  - difficulties of the rapid and extensive screening for pathogens.
- **Xenogeneic (different species) – fetal pig brain**. Transplantation of pig neurones into the striatum of a PD patient could potentially restore dopaminergic synaptic inputs onto neurones projecting out of the striatum, and allow resumption of normal movements. Experimental animal models have shown that neural xenografts, like neural allografts, can mediate behavioural recovery in a variety of neurodegenerative conditions. The major problem to a recipient of a neural xenogeneic cell preparation appears to be loss or rejection of the cells. Xenogeneic donor tissue is vigorously rejected, even in the presence of immune suppression. This hyperacute rejection depends on complement and preformed serum antibodies against xenogeneic antigenic determinants. Using transgenesis, pig tissues can be 'humanised' so that they are not rejected.

Gene therapy

The absence of effective long-term solutions for PD has led to the search for new gene therapy-based treatments.

PD gene therapy – the case for

- Good understanding of the pharmacology of the basal ganglia.
- Clinical proof that modulation of dopaminergic or glutamatergic pathways may provide symptomatic relief.
- The overall disability from the disease is severe.
- Current neurosurgical techniques are risky, but improving, and ablative procedures are providing increased information as to the functional anatomy of the basal ganglia.
- Well-defined animal models (Box 5.1).

PD gene therapy – the case against

- Tight control of L-DOPA concentrations is necessary, particularly late in disease progression. A symptomatic approach via dopaminergic modulation needs tight control of transgene expression.
- PD is chronic.
- PD is not acutely terminal.
- PD pathophysiology and progression are poorly understood.

Gene therapy strategies for the treatment of PD

The dopamine pathway

Genes encoding the enzymes of the dopamine biosynthetic pathway are potentially useful in gene therapy approaches to PD (Figure. 5.3b). A defective HSV vector has been used to deliver human tyrosine hydroxylase (TH) to the partially denervated striatum of the 6-hydroxydopamine lesioned rat. TH, the rate-limiting enzyme in the biosynthesis of catecholamines, is responsible for the conversion of tyrosine into L-DOPA (Figure. 5.3b). Behavioural and biochemical recovery was maintained one year after gene transfer (During *et al.*, 1994). Similarly, liposome-mediated gene transfer (Cao *et al.*, 1995) and AAV (Kaplitt *et al.*, 1994) have been used to introduce the TH gene into lesioned rats. In both cases, the apomorphine-induced rotational asymmetry of the lesioned rats was significantly reduced.

BOX 5.1: PARKINSON'S DISEASE ANIMAL MODEL

A common animal model for PD is the 6-hydroxydopamine lesioned rat. A stereotaxic injection of 6-hydroxydopamine is made into the substantia nigra. This selectively destroys dopaminergic cells, leaving postsynaptic neurones within the striatum intact, which experience up-regulation of their dopamine receptors. After two weeks, lesioned rats challenged with a dopamine agonist such as apomorphine display a forced rotation in the opposite direction to the side of the lesion. PD progression can be mimicked by monitoring the gradual degeneration of the substantia nigra dopaminergic neurones following an injection of 6-hydroxydopamine.

An alternative approach involved the infection of cultured, immature astrocytes, obtained from the prenatal rat brain, with a retrovirus carrying the human TH gene. DOPA-producing cells were then introduced into the striatum of 6-hydroxydopamine lesioned rats, with significant therapeutic efficacy.

Glutamate pathways

Glutamic acid decarboxylase (GAD) catalyses the conversion of glutamate to the inhibitory neurotransmitter gamma amino-butyric acid (GABA). The introduction of a GAD-expressing virus might have the effect of controlling the glutamatergic output of the subthalamic nucleus (Figure. 5.3a), thus offering symptomatic relief in advanced PD without the adverse effects of drugs. However, adverse effects of GABA are a potential hazard.

Neurotrophic factors

Neurotrophic factors **(http://www.euro.promega.com/pnotes/50/2883d/2883d.html)** that promote the survival of specific neuronal types during development, and following exposure to insults, have a potential role in the prevention of PD progression.

Glial cell line-derived neurotrophic factor (GDNF; **http://www.euro.promega.com/nnotes/nn007/5750a/5750a.html)** has been shown to exert trophic and protective effects on dopaminergic neurones. Both Ad (Bilang-Bleuel *et al.*, 1997; Choi-Lundberg *et al.*, 1997) and AAV (Mandel *et al.*, 1997) vectors have been used to transduce the regions of the rat brain close to the substantia nigra with the GDNF gene. Shortly after viral infection, the rats were subjected to 6-hydroxydopamaine lesioning of the substantia nigra. GDNF gene therapy was shown to protect dopamine neurones from progressive degeneration, and rotational asymmetry was significantly ameliorated.

5.9 Summary

- Genes can be efficiently introduced into the somatic cells of animals and humans.
- Viral vectors show great promise, both as tools for basic research and as tools for gene-mediated therapy.
- The absence of effective long-term solutions for Parkinson's disease has led to the search for new gene therapy-based treatments.
- Parkinson's disease-like symptoms in animal models have been effectively ameliorated following viral-mediated gene transfer.

Further reading

ESPEJO, E.F., MONTORO, R.J., ARMENGOL, J.A. and LOPEZ-BARNEO, J. (1998) Cellular and functional recovery of Parkinsonian rats after intrastriatal transplantation of carotid body cell aggregates. *Neurone* **20**: 197–206.

Autogeneic cell therapy for Parkinson's disease? Transplantation of dopaminergic glomus cells of the carotid bodies can restore motor function in a rat model of Parkinson's.

BRANCH, A.D. (1998) A good antisense molecule is hard to find. *Trends in Biochemical Science,* **23**, 45–50.

Critical discussion of the possible mechanisms by which so-called antisense oligonucleotides exert their effects.

FIRE, A., XU S.-Q., MONTGOMERY, M.K., KOSTAS, S.A., DRIVER, S.E. and MELLO C.C. (1998) Potent and specific genetic interference by double-stranded RNA in *Caenorhabditis elegans. Nature,* **391**, 806–811.

KREN, T.B., BANDYOPADHYAY, P. and STEER, C.J. (1998) *In vivo* site-directed mutagenesis of the factor IX gene by chimeric RNA/DNA oligonucleotides. *Nature Medicine,* **4**, 285–290.

Two remarkable papers that might herald the future of both transgenesis and gene therapy. Double-stranded nucleic acids, in one case double-stranded RNA, in the other an RNA/DNA chimera, are able to mediate chromosomal modification at high efficiency in vivo.

CHAPTER SIX

Visualising neuronal gene expression

Key topics

- Monitoring and measuring transcription
  - The nuclear run-on assay
- Monitoring and measuring RNA
  - Northern blotting
  - *In situ* hybridisation
  - RT-PCR
  - cDNA cloning
  - Sage
- Mapping transcripts
  - cDNA cloning and sequencing
  - RT–PCR and RACE
  - S1 mapping
  - Primer extension
  - RNase protection
- Monitoring and measuring protein
  - Radioimmunoassay
  - Western blotting
  - Immunocytochemistry
- Single cell gene expression assays
- Reporters of gene expression

6.1 Monitoring and measuring gene expression

Molecular neuroscience depends on our ability to monitor and measure the expression of genes. Techniques have been developed that enable gene products to be assayed, both qualitatively and quantitatively, at any point on the expression pathway – from **transcription** of the gene to the generation of the mature, functional peptide product. RNA products of specific genes are assayed by virtue of Watson and Crick base pairing between complementary probes or primers. Messenger RNAs are translated into protein products that can be detected by virtue of their carrying **epitopes** recognised by specific antibodies. Each of the methods described below has its advantages and its disadvantages, and the following questions need to be considered when either planning an experiment, or assessing data:

- What is the sensitivity of the technique?
- What is the anatomical resolution of the technique?
- Is the technique quantitative?
- Does the method tell us anything about the structure of the gene product?
- How can the results be interpreted in terms of gene function?

As you read, consider how the standard methods used in most molecular neuroscience laboratories to visualise gene expression measure up when tested with these questions. Also consider how these technologies will develop in the coming years. How might gene expression be monitored within the brain of a *living* organism? Could this ever be achieved non-invasively?

6.2 Monitoring and measuring transcription

The only method that enables transcription to be directly measured is the nuclear run-on assay (Box 6.1).

BOX 6.1: THE NUCLEAR RUN-ON ASSAY

The nuclear run-on assay (also known as the nuclear run-off assay) is a direct, accurate measure of the level of transcription of a gene. The nuclear run-on assay is used to measure differences in gene transcription quantitatively as a consequence of tissue-specific, developmental or physiological regulation. The method depends on the *in vitro* incorporation of radioactive ribonucleotides into RNA by RNA polymerase II complexes associated with nascent RNA chains within intact, isolated nuclei. The number of RNA polymerase complexes associated with a particular gene is a measure of the rate of transcription of that gene. Thus, the incorporation of radioactive ribonucleotides into a particular RNA is a measure of the rate of transcription. In this example, Gene A has more RNA polymerase complexes, and is being transcribed at a higher rate, than Gene B. The amount of label incorporated into Gene A nascent strands (shaded boxes) is greater than for Gene B. Transcription is assayed by isolating labelled RNA from nuclei following *in vitro* incubation with radioactive precursors. The radioactive RNA is used to probe

BOX 6.1: Contd

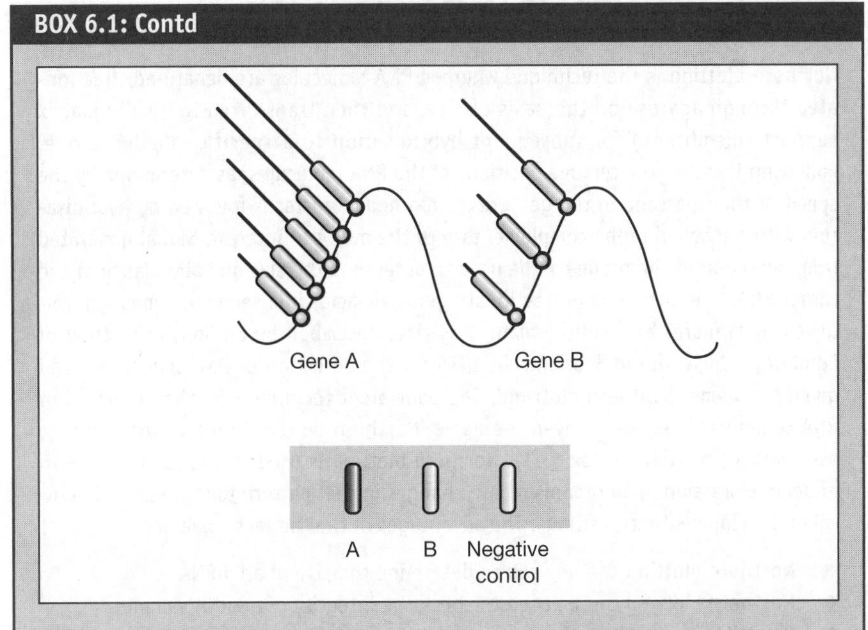

specific cloned DNAs fixed to a matrix. The level of hybridisation is directly related to the level of transcription of the gene of interest, and can be measured by scintillation counting or densitometric scanning following autoradiography.

6.3 Monitoring and measuring precursor RNA

Precursor RNAs are the nuclear intermediates between the primary transcribed RNA, and the mature message that is exported into the cytoplasm. Transcript processing intermediates can be detected in three ways:

1. By probing **Northern blots** (Box 6.2) of total cellular RNA with intron-specific probes.
2. By performing Northern blotting on RNA isolated from purified nuclei.
3. By *in situ* **hybridisation** (Box 6.3) using intron-specific probes.

In some cases, it has been shown that the level of a precursor mRNA correlates directly with the level of transcription of the gene that encodes it. However, this is by no means a hard-and-fast rule as precursor processing is itself a dynamic process that is subject to developmental and physiological regulation.

Why is RT–PCR (Box 6.4) an unsuitable method for examining precursor RNAs?

BOX 6.2: NORTHERN BLOTTING

Northern blotting is the technique whereby RNA molecules are denatured, fractionated through agarose on the basis of size, and then transferred to a solid matrix support (membrane) for subsequent hybridisation to a specific labelled probe. Following transfer, the relative positions of the RNA molecules, as determined by the speed of their passage in the gel, are maintained. Transfer is followed by hybridisation with a labelled probe complementary to the mRNA of interest. Signal generated from detection of the probe can be used to determine the size and abundance of the target RNA. The technique of fractionating nucleic acids in a semi-solid medium, followed by transfer to a solid matrix, was first described by Ed Southern, then of Edinburgh University in Scotland, in 1975 for the detection of DNA sequences, and given the name 'Southern blotting'. The equivalent technique for the detection of RNA sequences was hence given the name 'Northern blotting' (and as for proteins, see 'Western blotting' – Box 6.11). Northern blotting is used to analyse the pattern of gene expression in an organism, and changes in that pattern following physiological or developmental transitions. The advantages of the the technique are:

- Northern blotting can be used to determine the size of an mRNA.
- Northern blotting can be used to check the integrity of an RNA sample.
- Quantitation is straightforward and direct.
- Northern probes do not have to be completely homologous (can use 'cross-species' probes).
- By using probes corresponding to different exons, Northern blotting can be used to determine whether gene transcripts are subject to alternative splicing.
- By using probes corresponding to different 5' ends, Northern blotting can be used to determine whether a single gene encodes messages that are initiated at different sites.

BOX 6.3: *IN SITU* HYBRIDISATION

In situ hybridisation (ISH) has been used since the 1970s to localise specific mRNAs within morphologically preserved cells or tissues. Thin sections of the tissue are cut on a microtome or cryostat. These are probed with a labelled nucleic acid. Probes can be cDNA, oligonucleotide or RNA. Short oligonucleotide probes penetrate easily into tissue, and can be designed to distinguish between closely related transcripts. The use of a radioactively labelled probe is a proven approach for the detection of a tissue mRNA and is a good initial choice to assure accuracy. Following hybridisation, tissue sections on slides are washed, dried, and applied to X-ray film, which is exposed for a few days and developed. The sections can be stained and compared with the film. Alternatively, for more precise localisation, the slides can be dipped in photographic emulsion, exposed and then developed. The grains (signal resulting from exposure of the emulsion to the probe) are visualised in a layer just above the section. Both the tissue and the emulsion layer can be seen simultaneously through the light microscope. A major problem with

BOX 6.3: Contd

the use of radioactive probes is that the scatter inherent in isotope decay limits the spatial resolution of the technique. New methodologies for the generation and detection of non-radioactive probes **(http://biochem.boehringer-mannheim.com/prod_inf/manuals/InSitu/InSi_toc.htm)** have enabled the sensitive detection of mRNAs in tissue sections at high resolution. The extension of these techniques to the electron microscope has even enabled the direct visualisation of the precise subcellular location of specific mRNAs.

6.4 Monitoring and measuring mature RNA

The analysis of mature cytoplasmic RNA is often the easiest and quickest method of determining the pattern of gene expression in an organism, and how that pattern changes with developmental and physiological transitions. Steady-state RNA levels can be analysed by any one of a number of different techniques:

- Northern blotting (Box 6.2).
- *In situ* hybridisation (Box 6.3).
- RT-PCR (Box 6.4).
- cDNA cloning and sequencing (Chapter 1) and EST analysis (Chapters 1 and 2).
- Serial analysis of gene expression (Box 2.1).

BOX 6.4: RT–PCR

When a gene of interest is expressed in a small number of cells, or when transcripts are present in vanishingly low amounts, it is often difficult to detect transcripts using Northern blotting or *in situ* hybridisation. In these cases, the exquisite sensitivity of the polymerase chain reaction (PCR) can provide a solution. PCR was originally used to amplify target DNA sequences from complex mixtures, such as genomic DNA. This powerful technique has been modified to enable the amplification of specific RNA targets, usually derived from sequences expressed as mRNA. The first step in the RT–PCR protocol is to make a single-stranded complementary DNA (cDNA) copy of the RNA target. A primer is annealed to the RNA target and extended using the enzyme reverse transcriptase (RT). There are several options for first-strand primers – a specific sequence complementary to the target mRNA, an oligo dT primer that will anneal to the poly (A) tail at the 3' end of mRNA, or short random oligonucleotides that anneal to many internal locations along the length of the target RNA. The cDNA is then amplified exponentially by PCR using a thermostable DNA polymerase. The RT–PCR reaction

BOX 6.4: Contd

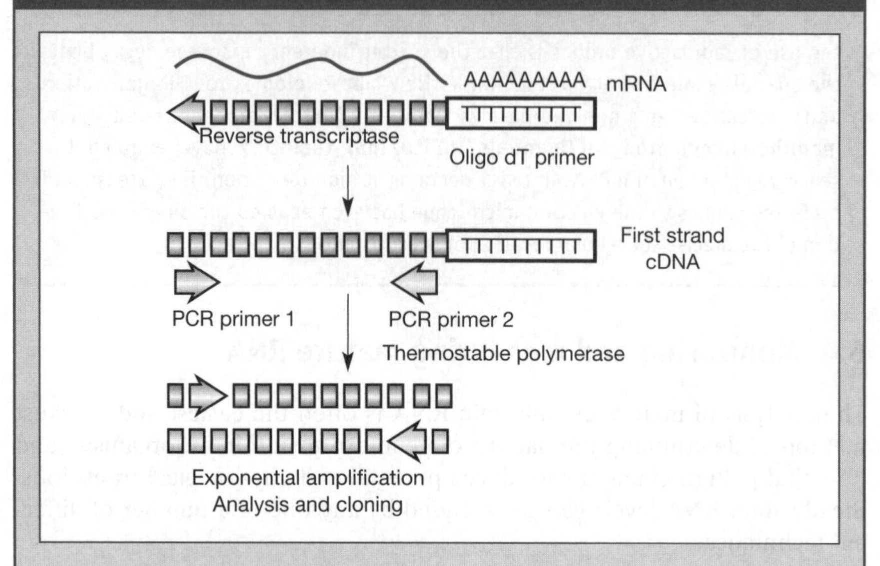

products can be visualised by fractionation on gels. Products can be cloned into plasmid vectors or directly sequenced.

RT–PCR is an extremely sensitive method for detecting the expression of a specific mRNA sequence in a tissue or cells. RT–PCR is also used to make cDNA clones. RT–PCR has been used to detect transcripts present at levels as low as 1 copy/1000 cells, and in samples as small as a single cell.

Competitive RT–PCR can be used to quantify the relative and absolute levels of an mRNA target. However, such quantitation requires very careful controls and attention to experimental design:

- Accurate determination of the absolute abundance of specific mRNAs requires the design of artificial constructs that are used to make RNA standards corresponding to the target RNA of interest.
- Accurate determination of relative abundance of specific mRNAs requires the judicious selection of an appropriate internal control RNA.
- For accurate quantification, both experimental and reference RNAs must be analysed within the linear range of the amplification reaction.

Advantages of RT–PCR

- Extremely sensitive.
- Detection is not limited by the amount of starting material.
- Fastest method of detection (less than one day).
- Large numbers of samples can easily be processed.
- Does not require specially constructed probes.
- Requires minimal sequence information.
- Primers can be designed to discriminate between closely related sequences that differ by as little as a single base.

Any differences in specific RNA levels observed in different tissues of an organism, or any changes in RNA level as a consequence of a physiological or developmental change, cannot, using these techniques, be ascribed to transcriptional controls. Such differences could equally be a consequence of post-transcriptional mechanisms that govern RNA stability. The steady-state level of a mature RNA is not a measure of the transcription rate of the gene that encodes it.

It is important to note that a positive hybridisation signal on a Northern blot, or in a tissue section following *in situ* hybridisation, cannot be interpreted as an indication of the presence of a functional, translatable messenger RNA. Structural analysis of RNAs is therefore necessary to determine if the transcript is authentic. The following methods allow RNA structures to be mapped:

- **cDNA** cloning and sequencing (Chapter 1)
- **RT–PCR** (Box 6.4) and **RACE** (Box 6.5)
- **S1 mapping** (Box 6.6)
- **Primer extension** (Box 6.7)
- **RNase protection** (Box 6.8)

BOX 6.5: RACE

RACE (Rapid Amplification of cDNA Ends) enables the amplification of unknown sequences at either the 5' end (5' RACE) or the 3' end (3' RACE) of a mRNA. RACE enables the identification of the ends of mRNA molecules, and allows full-length cDNAs to be derived. The only requirement is sufficient sequence information from the middle portion of the mRNA to design gene-specific primers (GSP).

5' RACE

First-strand cDNA is synthesised from total or poly(A) + RNA using the most 3' of the gene-specific primers (GSP1). Terminal deoxynucleotidyl transferase is then used to add a homopolymeric tail (usually A) to the 3' end of the cDNA, which corresponds to the 5' end of the mRNA. Tailed cDNA is then amplified by PCR using the second gene specific primer (GSP2) and an oligo dT-anchor primer. The cDNA can be further PCR-amplified using a nested, gene-specific primer (GSP3) and the PCR anchor primer. The 5' RACE products can then be cloned and sequenced.

3' RACE

3' RACE takes advantage of the poly (A) tail that is found at the 3' end of mRNAs. First-strand cDNA is synthesised from total or poly (A) + RNA using oligo dT as a primer. cDNA is then amplified by PCR using a gene-specific primer (GSP) and the oligo dT-anchor primer. The 5' RACE products are then cloned and sequenced.

BOX 6.5: Contd

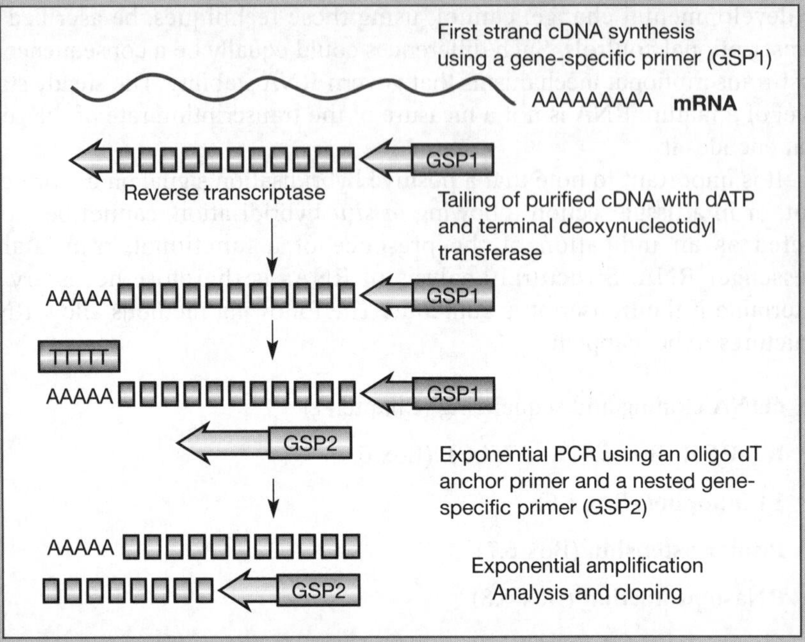

First strand cDNA synthesis using a gene-specific primer (GSP1)

AAAAAAAAA mRNA

GSP1

Reverse transcriptase

Tailing of purified cDNA with dATP and terminal deoxynucleotidyl transferase

AAAAA GSP1

TTTTT

AAAAA GSP1

GSP2

Exponential PCR using an oligo dT anchor primer and a nested gene-specific primer (GSP2)

AAAAA

GSP2

Exponential amplification
Analysis and cloning

(A) 5' RACE

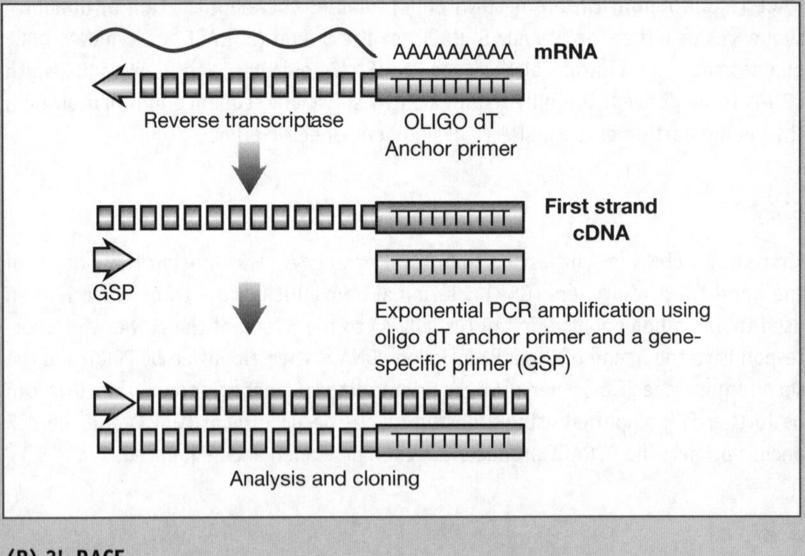

AAAAAAAAA mRNA

TTTTTTTTTT

Reverse transcriptase

OLIGO dT
Anchor primer

TTTTTTTTTT

First strand cDNA

TTTTTTTTTT

GSP

Exponential PCR amplification using oligo dT anchor primer and a gene-specific primer (GSP)

TTTTTTTTTT

Analysis and cloning

(B) 3' RACE

BOX 6.6: S1 NUCLEASE PROTECTION MAPPING

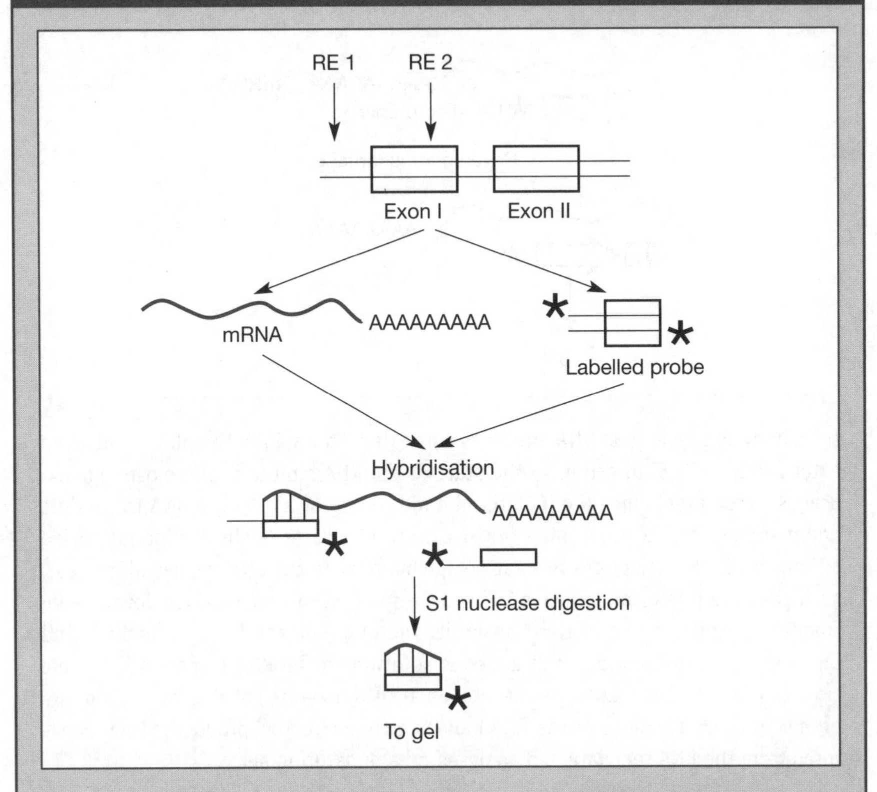

S1 nuclease mapping is used to map transcription start sites, transcription termination sites and intron–exon junctions. S1 nuclease is an endonuclease isolated from *Aspergillus oryzae* that digests single-, but not double-stranded, nucleic acid. It also digests partially mismatched double-stranded molecules with such a sensitivity that even a single base-pair mismatch can be cut, and hence detected. S1 nuclease mapping of a transcription initiation site requires a relatively detailed knowledge of the gene structure and sequence data (or a very good restriction map) of the first exon and several hundred bases of upstream sequence. When using S1 nuclease to map a transcription initiation site, the DNA probe normally consists of a radiolabelled genomic DNA fragment that starts within the first exon and extends upstream past the putative initiation site. In this example, the probe consists of a labelled genomic fragment isolated by using restriction enzymes (RE) 1 and 2. RE1 lies upstream of the putative start of transcription, whereas RE2 is within exon I. The DNA probe hybridises to the mRNA up to the first base transcribed; the upstream sequences remain unpaired. S1 nuclease will digest away the unpaired DNA and RNA, giving a protected fragment that corresponds in size to the length of the RNA from the transcriptional start site to the position of RE2. The product is analysed on a denaturing gel.

BOX 6.7: PRIMER EXTENSION

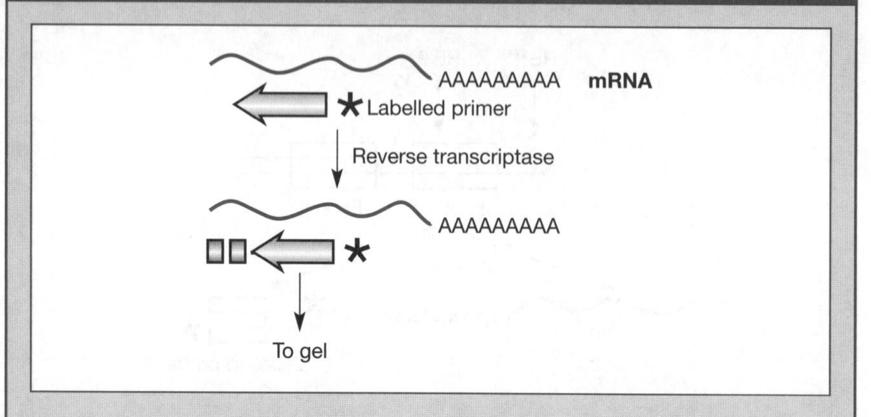

In primer extension, a DNA probe is annealed to an RNA template and then extended in a 3'- 5' direction to the start of the RNA molecule. The most common usages for this technique are in determining the size of full-length RNA transcripts and mapping transcription initiation sites. In most cases, a short oligonucleotide of length 25–35 nucleotides is made complementary to the RNA species of interest, such that the primer oligonucleotide will lie some 30–15 nucleotides downstream from the suspected end of the transcript. The oligonucleotide is normally 5'-end labelled and then hybridised in aqueous solution to the RNA transcript. Reverse transcriptase is then used in the presence of deoxynucleotides to extend the primer to the 5' terminus of the RNA molecule. The extended product is then denatured from the RNA template and analysed on a denaturing gel.

BOX 6.8: RNASE PROTECTION

The RNase protection assay is based on the resistance of hybridised RNA to single-strand-specific RNases. An RNA strand, in a duplex with either DNA or another RNA strand, will not be degraded by either RNase A or RNase T1. RNAse protection is used to map the ends of RNA molecules, or exon–intron boundaries. It also provides an attractive and highly sensitive alternative to Northern blot hybridisation for the quantitative determination of mRNA abundance. The probe fragment is cloned into a transcription vector under the control of a bacteriophage promoter (either the T3, T7 or SP6 promoter). Using the corresponding polymerase, a radiolabelled RNA transcript of high specific activity is generated. Hybridisation is carried out with an excess concentration of probe so that all complementary sequences within the target RNA are driven into the labelled hybrid. Unhybridised probe, or any single-stranded regions of the hybridised probe, are then removed by RNase digestion. The 'protected' probe is fractionated on a denaturing gel, and exposed to film for visualisation and quantitation.

BOX 6.8: Contd

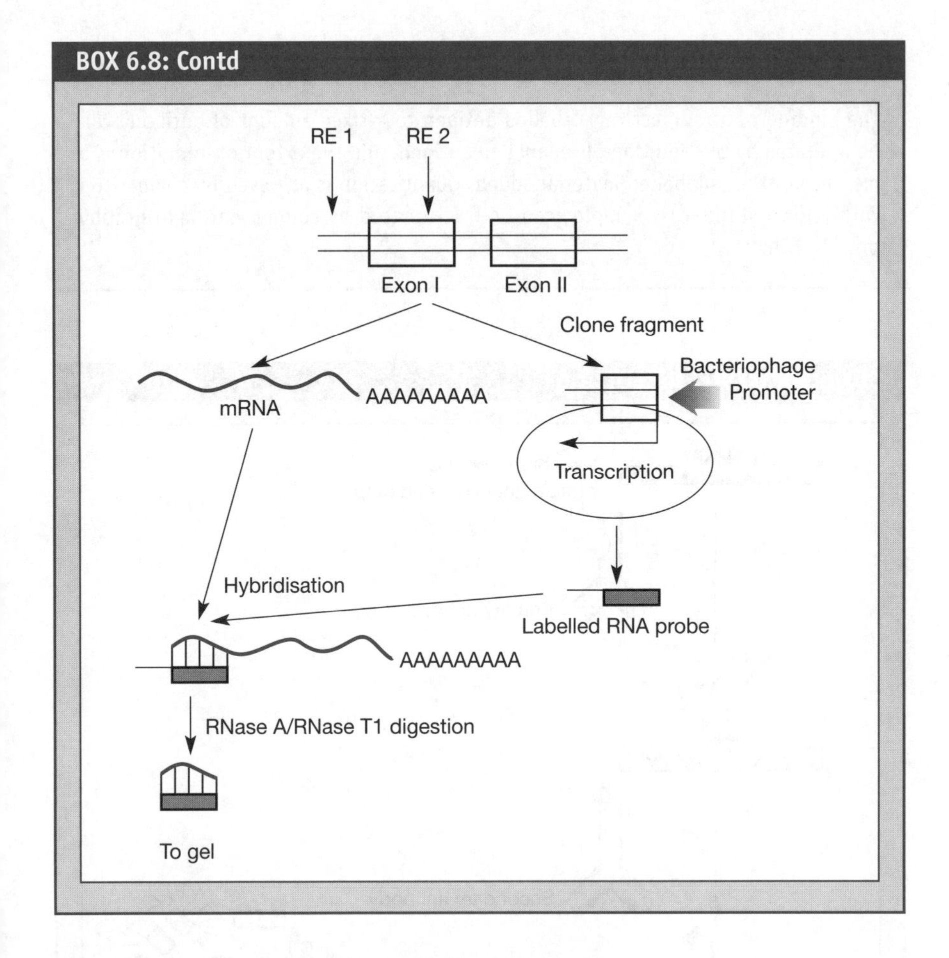

6.5 Measuring and monitoring protein

The presence of an authentic RNA in a cell is not necessarily indicative of peptide synthesis. Structurally authentic mRNAs may be poorly handled by the translational machinery in a cell. Even the efficient translation of a functional message may not result in biologically active peptide if post-translational processes are absent, or if the peptide is rapidly degraded. The analysis of gene expression must therefore extend to the description of the protein end-points of the expression pathway. Included in this must be a consideration of protein function. The following methods are commonly used to study the expression of proteins in the brain:

- Radioimmunoassay (Box 6.9)
- **Western blotting** (Box 6.10)
- Immunocytochemistry (Box 6.11)

BOX 6.9: RADIOIMMUNOASSAY (RIA)

The binding of a radioactively labelled antigen to a fixed amount of antibody can be inhibited by the addition of unlabelled antigen, and the extent of inhibition is a measure of the unlabelled material added. Quantitation is achieved by comparison with known standards. Samples can be fractionated by column chromatography prior to RIA.

BOX 6.10: WESTERN BLOTTING

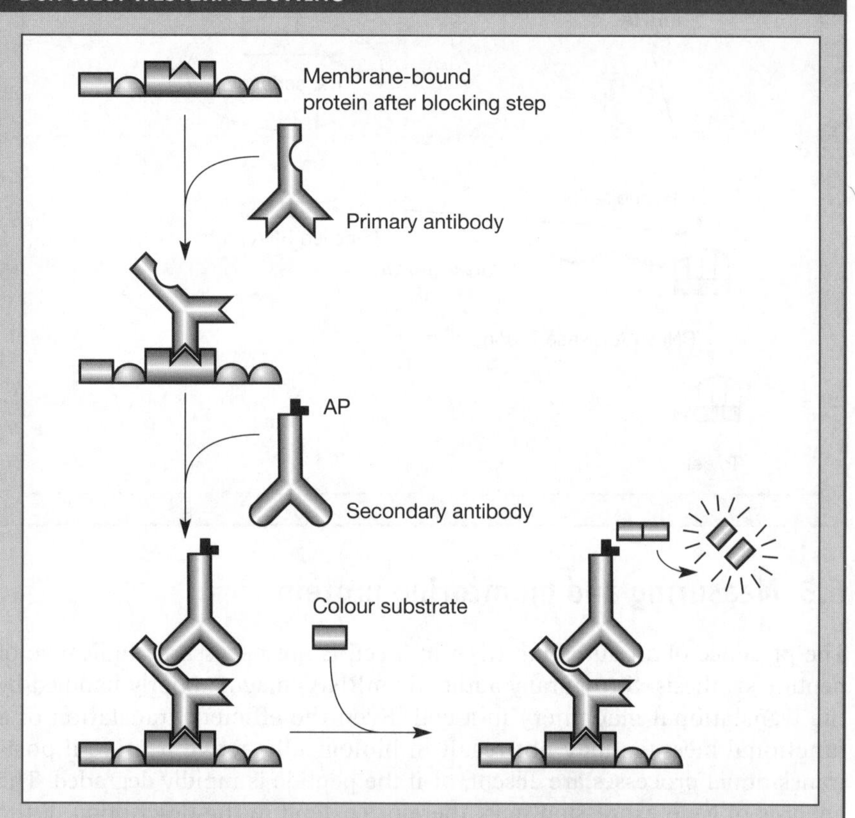

Western blotting enables the detection of specific proteins on the basis of size following electrophoretic separation. Protein extracts are fractionated in sodium dodecyl sulphate (SDS)–polyacrylamide gels. The protein sample is denatured and coated in detergent by heating in the presence of SDS and a reducing agent. The SDS coating gives the protein a negative charge that is proportional to its length. The negatively charged proteins migrate towards the anode when a voltage is applied accross the gel. The gel matrix acts as a molecular sieve, and proteins are separated solely on the basis of their mass. The fractionated pro-

BOX 6.10: Contd

teins are then transferred electophoretically to a solid matrix, usually a membrane made of nitrocellulose. Antigens immobilised on membranes are detected with antibodies in a three-step process. Firstly, the primary antibody, an IgG directed against the antigen in question, is incubated with the membrane. In the second step, a secondary antibody which recognises all IgGs (anti-IgG), is added to find locations on the membrane where the primary antibody has bound. The secondary antibody is conjugated to an active moiety that enables its detection – for example, the secondary antibody might be linked to the enzyme alkaline phosphatase (AP). Thus, in the third step, the membrane is incubated with an appropriate enzyme substrate, resulting in the deposition of a coloured product on the membrane. This coloured product provides a visual read-out of primary antibody recognition of a specific protein.

BOX 6.11: IMMUNOCYTOCHEMISTRY

Immunocytochemistry is used to localise the peptide product of a gene, or to identify particular cell types within tissue sections by labelling a specific protein. Tissue sections are first incubated with a primary antibody, an IgG that specifically recognises an epitope or epitopes on the protein of interest. The sections are then incubated with a secondary antibody which recognises all IgGs (anti-IgG), and is added to find locations within the tissue section where the primary antibody has bound. The secondary antibody is conjugated to an active moiety that enables its detection – the secondary antibody might be linked to:

- an enzyme, such as alkaline phosphatase or horseradish peroxidase. To detect an enzyme-linked secondary antibody, the tissue section is incubated with an appropriate enzyme substrate, resulting in the deposition of a coloured product. This coloured product provides a visual read-out of primary antibody recognition of a specific protein;
- a fluorescent tag, which is visualised by viewing it through a microscope equipped with an ultraviolet light source that excites the molecules of the tag, resulting in the emission of visible light;
- a gold particle. Tiny gold particles can be viewed under the electron microscope, enabling the precise subcellular localisation of specific proteins.

6.6 Co-localisation

It is often important to ask if particular cells within the CNS express two or more genes of interest. This can be achieved by: combining *in situ* hybridisation and immunocytochemistry; using two or more probes with different labels in the same immunocytochemistry or *in situ* hybridisation experiment. For example, different oligonucleotide probes can be simultaneously detected if one is radioactively labelled, and the other is non-radioactively labelled.

6.7 Single cell measurements

The patch–clamp technique can be combined with RT–PCR (Box 6.4) to correlate the expression of gene transcripts with functional properties of specific individual neurones in the brain. Cells in brain slices can be identified on the basis of their morphological, immunocytochemical and electrophysiological characteristics. Even better, transgenic animals expressing fluorescent reporters in specific neuronal cell types can be used to visualise directly and identify that cell type under the microscope. After electrophysiological recording, the contents of the cell can be harvested using the same patch pipette. RT–PCR can then be used to study the expression of specific genes in that single cell, or to generate single cell cDNA libraries (see section 1.5).

6.8 Reporter genes

Using transgenic technologies (see Chapters 4 and 5), new genes can be placed into the brain, and this presents us with new, more sensitive ways of monitoring and measuring gene expression. Transgenes can be constructed so that they incorporate **reporter genes** – genes that code for proteins that have a unique enzymatic activity. The use of reporter genes in transgenic animals provides a rapid method for the detection of transgene expression, which is easily distinguishable from expression of the corresponding endogenous gene of the animal. The regulatory sequences of a chosen gene are fused to a readily assayable protein coding region, for example:

- Chloramphenicol acetyl transferase (CAT; 6.8.1)
- β-galactosidase (6.8.2)
- Luciferase (6.8.3)
- Green fluorescent protein (6.8.4).

Sensitive assays are available for each of these proteins that facilitate detection and quantitation of transgene expression.

6.8.1 Chloramphenicol acetyl transferase

CAT is a bacterial enzyme with no mammalian counterpart. The enzyme catalyses the following reaction:

Chloramphenicol + acetyl CoA → 3-acetylchloramphenicol + CoA

The CAT assay is based on the acetylation of ^{14}C-chloramphenicol using acetyl CoA as the acetyl donor group. The acetylated products of this reaction are then separated from the non-acetylated substrate by thin-layer chromatography and analysed by autoradiography. This process of separation is time-consuming and has led to the development of a non-chromatographic assay for CAT activity that uses an organic solvent to extract both the acetylated and non-acetylated forms of chloramphenicol from labelled acetyl CoA.

6.8.2 β-galactosidase

The enzyme β-galactosidase, which catalyses the hydrolysis of β-galactosides, including lactose, is encoded by the *E. coli LacZ* gene. Enzyme activity is measured by a simple photometric assay that measures the hydrolysis of the substrate *o*-nitrophenyl β–D-galactopyranoside (ONPG). β-galactosidase can also be monitored histochemically using the substrate X-gal (5-bromo-4-chloro-3-indoyl β-D-galactoside).

6.8.3 Luciferase

The reporter gene for luciferase was cloned from the firefly (*Photinus pyralis*), a bioluminescent organism. Firefly luciferase catalyses the following reaction:

$$Luciferin + ATP + O_2 \rightarrow oxyluciferin + PP_i + CO_2 + light + AMP$$

The expression of firefly luciferase is measured in the presence of ATP and luciferin by the emission of light, which is measured using a luminometer. The assay is very sensitive and may be used in the analysis of the transcriptional activity of weak promoters. The luciferase assay also has the advantage of using a non-radioactive substrate, and there is no similar enzyme activity in mammalian cells.

6.8.4 Green fluorescent protein

Green Fluorescent Protein (GFP) from the jellyfish *Aequorea victoria* enables gene expression and protein localisation *in vivo*, *in situ* and in real time. GFP emits bright green light upon exposure to ultraviolet light. Unlike other bioluminescent reporters, GFP does not require any additional proteins, substrates or cofactors. GFP fluorescence is stable, species independent, and can be monitored non-invasively in living cells (**http://www.clontech.com/clontech/Catalog/GeneExpression/GFPintro.html**) Quantitation of expression can be achieved using a fluorimeter (cell extracts), confocal microscopy or fluorescence activated cell sorting (FACS) analysis.

A video clip of GFP fluorescence in a living cell has recently been published as a multimedia adjunct to a *Biochemical Journal* paper, and can be viewed at: **http://bj.portlandpress.co.uk/bj/333/bj3330193add3.htm**

6.9 Summary

- The detection and quantitation of specific nucleic acid gene products depends upon Watson and Crick base pairing between probe and target.
- The detection and quantitation of specific proteins depends upon antibody recognition of unique epitopes.
- Methods differ in terms of:
  - sensitivity
  - resolution
  - ease of quantitation
  - the availability of structural information
- Sensitive techniques enable gene expression to be examined in single neurones.
- Reporters can be incorporated into transgenes that allow expression to be easily monitored and quantitated.
- The reporter Green Fluorescent Protein enables gene expression to be observed in living cells in real time.

CHAPTER SEVEN

Plasticity of neuronal gene expression

Key topics

- Physiological regulation
- Vasopressin gene expression
  - Changes in mRNA level and size
  - Transcriptional and post-transcriptional
  - Cell-specific mechanisms
- Neurotransmitter receptor gene expression
- Cellular immediate-early gene expression
  - AP-1 genes – fos and jun
- Integration of nuclear signalling in neurones

7.1 Introduction

Following the refinement of ***in situ* hybridisation** technologies in the early 1980s (see Box.6.4), studies of gene expression in the adult brain showed that mRNA expression could be dramatically altered by physiological stimuli. Thus, neuronal gene expression was shown to exhibit **plasticity** – fully differentiated neurones retain the capacity to undergo phenotypic changes that may serve as an adaptive response to functional demand. Such changes in mRNA expression would commonly be expected to be transcriptionally mediated, and subsequent studies of **inducible transcription factor** expression in the brain, begun in the late 1980s, also revealed stimulus-dependent changes in expression. Evidence of post-transcriptional mechanisms of mRNA regulation has also been obtained. This chapter will describe studies of genetic plasticity in the brain, and show how the field is advancing to provide insights into pathophysiological mechanisms.

7.2 Physiological regulation of neuronal gene expression

The capacity of neuronal systems to adapt following a change in functional demand can involve both non-genomic and genomic mechanisms. The same is true of the changes in synaptic strength that appear to be related to learning and memory (see section 8.2.3) Adaptive genomic responses have been observed in many neuronal systems, and can be illustrated with reference to the neuropeptide gene that encodes arginine–vasopressin (AVP).

7.2.1 Regulation of vasopressin gene expression

Aside from interest in the neuroendocrine role of AVP, the hypothalamic AVP system has been selected as a model for studies of neuronal gene plasticity because:

- a variety of simple, function-related stimuli elicit changes in AVP gene expression;
- cell-type-specific physiological responses are observed:
    - magnocellular neurones (e.g. supraoptic nucleus) respond to osmotic stimuli (e.g. dehydration)
    - parvocellular neurones (e.g. parvocellular paraventricular nucleus) respond to stress stimuli
    - parvocellular neurones (suprachiasmatic nucleus) respond to circadian stimuli
- the gene is small, therefore facilitating the construction of transgenes (see Box 4.2).

Regulation of mRNA level and size

As detailed in Chapter 6, a variety of techniques can be used to visualise mRNA expression. The importance of using a variety of approaches has been illustrated through the study of AVP mRNA regulation where different insights have been gained from alternative techniques. For example, *in situ* hybridisation histochemistry was initially used to demonstrate that the abundance of AVP mRNA is increased in magnocellular hypothalamic neurones following an osmotic stimulus, but it was not until further analysis using Northern blots (see Box 6.2) was carried out that a stimulus-dependent change in the *size* (as determined by relative migration on agarose gels) of AVP mRNA was also found (Figure 7.1a). Such a difference in mRNA size could possibly result from either:

- Use of an alternative transcription initiation site.
- Alternative splicing of the RNA precursor, involving an alternative or additional exon.
- Alternative transcriptional termination, involving a distal polyadenylation site.
- A change in the length of the mRNA poly (A) tail.

The first three of these alternatives could affect the coding potential of the mRNA and possibly protein function. Unknown transcripts are typically characterised using a variety of analytical methods including **RNAse protection, primer extension** (see Sambrook *et al.*, 1989) and **RACE** (**r**apid **a**mplification of **cDNA e**nds; see Box.6.5). In the case of the larger, osmotic stimulus-dependent AVP mRNA, analysis revealed that the transcript sequence was identical to that of mRNA extracted from control animals – the size difference was due to an increase in poly (A) tail length of approximately 150 bases (Figure 7.1a). Physiological changes in mRNA poly (A) tail length have now been recognised for a number of neuronal genes; the function of this modification in mRNA structure in the adult is undefined but evidence from developmental systems suggests that an extended (A) tail may relate to either enhanced mRNA stability, or enhanced translation. Clearly, either of these possibilities could contribute to an increased biosynthesis of the AVP precursor polypeptide which is required for adaptation to the change in physiological demand for AVP during the period of an osmotic stimulus.

Transcriptional versus post-transcriptional mechanisms

The regulation of mRNA poly (A) tail length described in the previous section suggests that the observed plasticity in AVP mRNA expression during an osmotic stimulus may be post-transcriptionally mediated, at least in part – mRNA levels would accumulate by virtue of enhanced stability, and therefore reduced turnover. It is not always appreciated, in fact, that a change in the level of mRNA does not necessarily reflect a change in either the initiation or rate of transcription. In the case of osmotically stimulated AVP mRNA accumulation, however, clear evidence of a transcriptional mechanism of up-regulation was obtained initially using **nuclear run-on**

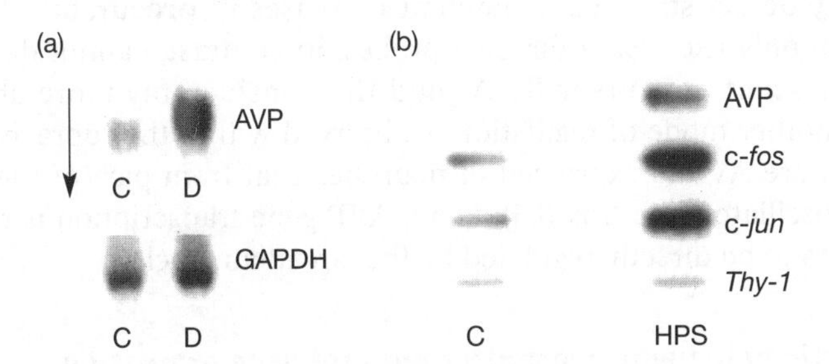

Figure 7.1 Regulation of vasopressin (AVP) gene expression. (**a**). Northern blots of hypothalamic RNA showing an increase in both the level and size (as determined by gel migration, in direction of arrow) of AVP mRNA in dehydrated rats (D) as compared with controls (C). Note that GAPDH mRNA, measured as a control, does not exhibit similar changes. (**b**). Nuclear run-on assay of AVP gene transcription in the hypothalamus showing an induction of transcription following a hypertonic saline injection (HPS). Note that similar changes are observed for the immediate-early genes c-*fos* and c-*jun*, but the rate of *Thy-1* gene transcription does not change.

analysis of transcription (see Box.6.1) which provides a direct measure of transcription (Figure 7.1b.). The increase in transcription was found to be sufficient to account fully for the increase in mRNA level; therefore the proposed post-transcriptional contribution to mRNA accumulation appears to be of minor importance in this context. An alternative approach to monitoring transcriptional up-regulation *in vivo* using intron (rather than exon) specific probes has also been applied to the study of AVP gene regulation (see next section). Furthermore, a recombinant AVP promoter has been shown to mediate osmotic up-regulation of a reporter gene in transgenic rats (Zeng *et al.*, 1994).

Post-transcriptional regulation has been revealed in studies of AVP gene expression in a different neuronal system. AVP neurones in the bed nucleus of the stria terminalis (BNST; see Figure 7.2.) differ from hypothalamic neurones in their co-expression of gonadal steroid hormone receptors. Accordingly, the AVP gene in the BNST is subject to regulation by gonadal steroids – there is a marked sex difference in the expression of the gene in this region, and castration followed by testosterone replacement in males results in reciprocal changes in BNST AVP mRNA levels as determined by Northern analysis. However, nuclear run-on analysis of transcription does not reflect this change, and it is apparent in these neurones that a post-transcriptional mechanism may be operating.

Cell-specific gene regulation

In addition to the cell-type-specific regulation of the AVP gene within the BNST (see previous section), specific modes of regulation are also observed within other neuronal subgroups. In parvocellular neurones of the hypothalamic paraventricular nucleus, AVP mRNA is selectively up-regulated by stress – a transcriptional mechanism of regulation has been indicated by the use of **intron-directed** *in situ* hybridisation probes (Herman, 1995) which specifically detect stimulus-dependent increases in **precursor RNA.** The more commonly used exon-directed probes, in contrast, cannot distinguish between nascent, precursor RNAs and the considerably more abundant mRNA. Another mode of regulation is observed within the suprachiasmatic nucleus where AVP is expressed in neurones that from part of the central circadian oscillator (see Box 8.3). Here, AVP gene transcription is rhythmic and appears to be directly regulated by the circadian clock.

7.2.2 Plasticity in neurotransmitter receptor gene expression

Studies of a gene closely related to AVP, oxytocin, have also provided evidence of physiological regulation of neuropeptide gene expression (see Box p. 109). Additionally, the oxytocin system has been exploited to demonstrate a physiological switch in neurotransmitter receptor subunit expression (Brussaard *et al.*, 1997). In this study a reduction in the GABAergic inhibition of oxytocin neurones during late pregnancy has been

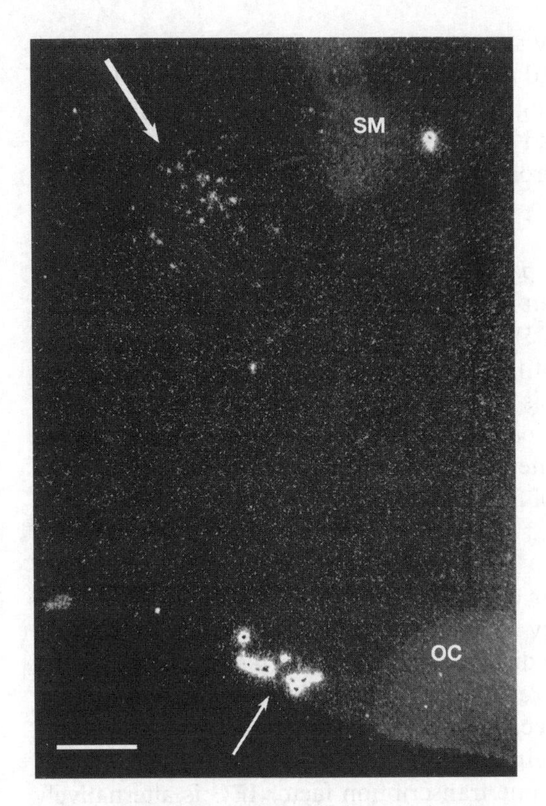

Figure 7.2 Vasopressin mRNA in neurones of the supraoptic nucleus (SON, small arrow) and bed nucleus of the stria terminalis (BNST, large arrow) in the mouse forebrain. The mRNA has been visualised using *in situ* hybridisation (see Box 6.3) – here a 35S-labelled probe has been hybridised to vasopressin mRNA, and detected using a photographic emulsion. Note that the SON neurones express AVP mRNA at a much higher level than the BNST neurones. The brain landmarks shown are OC, optic chiasm, and SM, stria medullaris. Scale bar: 50 μm. Reprinted from Smith, M. and Carter, D.A., *In situ* hybridisation analysis of vasopressin mRNA expression in the mouse hypothalamus: diurnal variation in the suprachiasmatic nucleus, *Journal of Chemical Neuroanatomy*, **12**, 105-112. Copyright 1997, with permission from Elsevier Science.

shown to correlate with a decrease in the α1: α2 $GABA_A$ receptor mRNA subunit ratio. Further evidence of receptor expression plasticity in the adult brain has been obtained in studies of glutamate receptor subunit mRNA changes following corticosterone treatment (see Nair *et al.*, 1998).

7.3 Cellular immediate-early gene expression

Studies in the early 1980s showed that c-*fos* and other cellular counterparts of viral oncogenes were rapidly and transiently induced in cultured neurones in response to growth factors and neurotransmitters (see Morgan and Curran, 1991; transient transcriptional induction can be viewed at: **http://www.fas.harvard.edu/~ashaywit/Fos.html).** These *in vitro* experiments precipitated a new wave of molecular neuroscience research because it was soon discovered that c-*fos* could be induced in the brain of experimental animals following a variety of physiological and pharmacological stimuli. The widespread application of this experimental approach across the breadth of the neuroscience research community was based on two premises:

1. The expression of c-*fos* in particular neurones reflects stimulus-specific cellular activation, and can therefore be used to map functional pathways in the brain (Figure 7.3).
2. The induced expression of transcription factors like the c-Fos protein forms the mechanistic basis for plasticity in neuronal gene expression. Thus, c-Fos acts as a nuclear messenger or 'molecular switch' which links short-term second messenger signals to long-term responses by altering the transcription of functional 'target genes'.

Many other genes apart from c-*fos* exhibit the characteristic transient induction that has led to their classification as **cellular immediate-early genes** (IEGs). However, many of the cellular IEGs are not transcription factor genes but encode quite different cellular molecules including secretory proteins and membrane receptors. Further, the nomenclature of the cellular IEGs is confusing because many of the genes have alternative names which derive from their simultaneous identification by different research groups; for example, one transcription factor IEG is alternatively referred to as NGFI-A, krox-24, zif-268, EGR-1 and TIS8. A computer search of publications on neuronal IEG expression (e.g. using a search tool such as Medline) will reveal a formidable volume of experimental data but, as discussed below, the functional relevance of these phenomena is mostly undefined (see Guest Box p. 109).

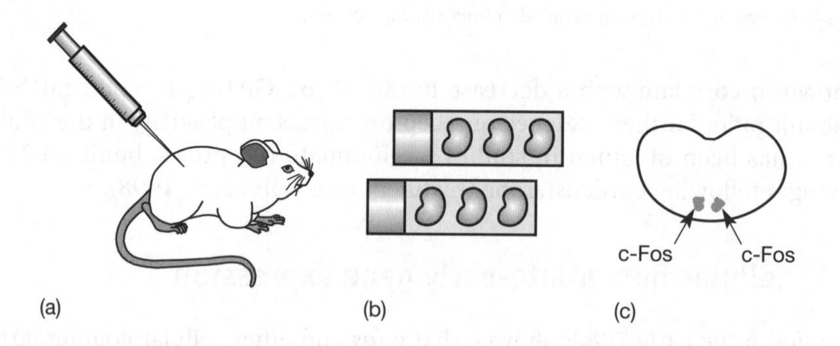

Figure 7.3 Mapping of functional activity in the brain by c-Fos immunohistochemistry. (**a**) The experimental paradigm, such as injection of a particular neuroactive drug, is performed with careful attention to control procedures. (**b**) Multiple brain sections are obtained using standard histological procedures. (**c**) Activated neurones are identified using c-Fos immunohistochemistry.

GUEST BOX BY G. LENG

Synaptic regulation of c-*fos* expression.

When I began as a neuroendocrinologist, I was concerned mainly with how secretion was controlled – what stimuli were involved, and how was the information encoded in neural pathways. However, about 10 years ago I and many others began to appreciate the importance of synthesis. Many neurosecretory neurones synthesise several secretory products – oxytocin cells, my particular favourite, also make the opioid peptide dynorphin. We had worked out that opioid peptides were co-secreted with oxytocin, and acted back on the neurosecretory nerve terminals to inhibit secretion. This finding seemed clear, but did not make particular sense – if the object was to restrain oxytocin secretion, why not just make less? However, if the *synthesis* of opioids and oxytocin was regulated *differentially*, then the releasability of oxytocin might vary in different physiological circumstances according to how much opioid was produced.

Oxytocin synthesis is elevated after an increase in secretion, apparently to replenish the stores, but what exactly triggers the increase in synthesis? A possible answer was suggested by evidence that neurosecretory activation was accompanied by the expression of immediate-early genes, including c-*fos*. These genes encode putative transcription factors, and their expression is induced in response to second messenger activation, to increased intracellular calcium concentrations, and possibly as a direct response to membrane depolarisation. Pretty well every stimulus we tested which activated oxytocin secretion also induced expression of c-*fos* in oxytocin cells. Indeed, we found increased c-*fos* mRNA expression in the supraoptic nucleus just 10 minutes after systemic injection of cholecystokinin – a stimulus which increases oxytocin cell firing rate by just 1 spike per second. Even more satisfyingly, we found Fos, the protein product of c-*fos*, in oxytocin cells during parturition, showing that the expression of c-*fos* was part of the physiological repertoire of oxytocin cells.

I had persuaded myself that c-*fos* was induced as a direct consequence of electrical activation, but Simon Luckman, then a post-doctoral researcher in my laboratory, proved me wrong. The axons of oxytocin cells all project to the pituitary; electrical stimuli applied here evoke action potentials that are propagated back along the axons to invade the cell bodies – mimicking the electrical excitation which follows natural stimuli, but without activating afferent pathways to the oxytocin cells. Simon showed that pituitary stalk stimulation released oxytocin, but did *not* induce c-*fos* expression. So, induction of c-*fos* in oxytocin neurones is a consequence of afferent (i.e. synaptic) activation, and not a consequence of electrical activation *per se*.

It is not clear that c-*fos* has any role in regulating oxytocin expression, but it may regulate dynorphin expression. During pregnancy, the oxytocin content of the

GUEST BOX Contd

pituitary increases markedly towards term, in preparation for parturition. The increase probably results not from increased synthesis, but from increased restraint of secretion – from a functional up-regulation of the dynorphin-mediated autoinhibition. This up-regulation coincides in mid pregnancy with the secretion of high concentrations of relaxin from the corpus luteum in the ovary. Relaxin activates oxytocin cells via an afferent pathway from the subfornical organ, a part of the brain which lacks an effective blood–brain barrier; does this activation induce a transient expression of c-*fos*, leading to a transient up-regulation of co-expressed dynorphin? This is one of many things we do not yet know.

LUCKMAN, S.M., DYBALL, R.E.J. and LENG, G. (1994) Induction of c-*fos* expression in hypothalamic neurons requires synaptic activation and not simply increased spike activity. *Journal of Neuroscience*, **14**, 4825–4830

After completing his PhD, **Gareth Leng** joined The Babraham Institute in Cambridge as a scientist in the research group of Barry Cross, a pioneer in neuroendocrinology, then director of the Institute. He stayed at Babraham for 17 years, working mainly on the neural regulation of oxytocin and vasopressin release, before moving to the Chair of Experimental Physiology at The University of Edinburgh where today he heads a large research group in Neuroendocrinology.

Regulation of the AP-1 family of IEGs – Fos and Jun

Most IEG studies have focused on the AP-1 family of genes – so-called because it includes one member originally termed **a**ctivator **p**rotein-**1** which binds to a cognate *cis*-acting element also known as AP-1. There are eight AP-1 proteins, related to either c-Fos or c-Jun:

Fos-related	*Jun-related*
c-Fos	c-Jun
Fra-1	JunB
Fra-2	JunD
Fos-B	
ΔFosB	

This group of IEGs is useful to consider because they exhibit a variety of functional characteristics that are illustrative of other similar factors. The wide availability of both mRNA probes and antibodies has also facilitated extensive studies of AP-1 expression that have revealed both cell- and stimulus-specific patterns of regulation.

The use of c-*fos in situ* hybridisation histochemistry or c-Fos immunocytochemistry as markers of neuronal activation has now gained general acceptance. A recent application of this approach has enabled the identification of a novel pattern of ventrolateral preoptic neuronal activation during sleep (Sherin *et al.*, 1996). Rigorous attention to experimental controls is required, however – one early study showed that simply moving a cage of animals to an adjacent laboratory resulted in c-*fos* induction in certain neuronal systems.

Functional roles of Fos and Jun proteins in the brain

Experimental studies have indicated functional roles for Fos and Jun proteins in numerous cellular and system-level mechanisms including:

- neuronal death (see Kasof *et al.*, 1995);
- pain (see Herdegan and Zimmermann, 1995);
- neurotrophic factor signalling (Segal and Greenberg, 1996; also see **http://www.fas.harvard.edu/~ashaywit/Neurones.html)**;
- learning and memory (see Draganow, 1996);

A required functional correlate of an increase in AP-1 transcription factor expression is an increase in AP-1 DNA binding activity. This is readily measured using a **gel retardation ('band-shift') assay** and dramatic changes in DNA binding activity can be measured in the hippocampus, for example, following treatment with an excitotoxin such as kainic acid. The composition of the AP-1 binding complex can then be determined using different protein-specific antisera in a **'supershift'** assay (Figure 7.4.).

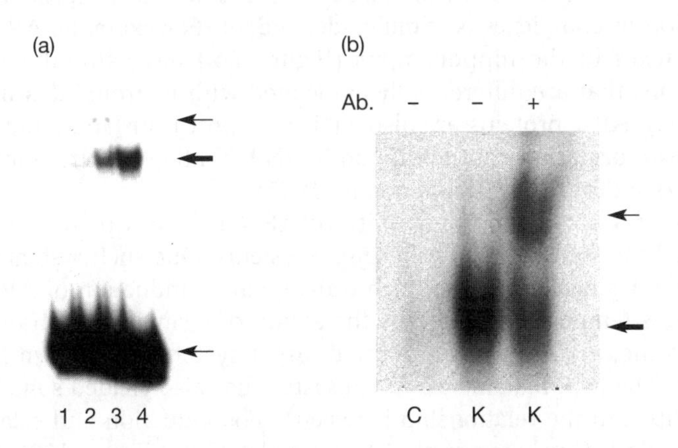

Figure 7.4 Regulation of AP-1 DNA binding activity detected with band shift assay. (a) Example of electrophoretic mobility shift ('gel shift' or 'band shift') assay showing the 'shift' (slower migration) of radioactive probe in the presence of increasing amounts (1→4) of AP-1 protein (large arrow). The top of the gel, and the 'free' (unbound) probe are shown by small arrows. **(b)** Detail of a band shift assay showing the induction of AP-1 activity in the hippocampus by an injection of kainic acid (K) as compared with control tissue (C). In the presence of an antibody to c-Fos protein (+), migration of the AP-1 band (large arrow) is slowed further by the formation of a ternary (probe–protein–antibody) complex (small arrow).

Despite the extensive evidence of AP-1 regulation in neural systems, the proposed role of Fos and Jun proteins as nuclear messengers that link neuronal activation to changes in 'functional' gene expression (section 7.3) has received limited experimental support. Null mutations, of the c-*fos* gene, for example (Wang *et al.*, 1992), are not associated with marked deficits in neuronal function; such findings may however simply indicate redundancy within the AP-1 (and related gene) family, and developmental reassignment of the c-Fos functional repertoire in the null mutants. The technical difficulty of selectively blocking the activity of specific regulatory proteins in neuronal subgroups has been an obstacle, although recent advances in gene targeting (see Chapters 4 and 5) have now facilitated such experiments. Stereotaxic injections of antisense *fos* and *jun* oligonucleotides is another approach that has also been employed with some success (Wollnick *et al.*, 1995).

It is important to remember that inducible transcription factors like c-Fos and c-Jun do not act in isolation – the potential functional effects of neuronal AP-1 expression must be considered in the context of the multitude of interacting transcriptional regulatory molecules which contribute to the overall integration of nuclear signalling. Such considerations have both qualitative and quantitative aspects – it has been recognised for some time that the relative amounts of Fos and Jun proteins can specify either positive or negative actions of other transcription factors (Diamond *et al.*, 1990). Certain AP-1 factors including JunB and Fra-2 can also exert inhibitory actions at AP-1 promoter elements by attenuating the activity of positively acting factors. Clearly, therefore, the composition of the AP-1 factor complex, which can involve a multitude of different Fos and Jun combinations is an important determinant of function, and evidence suggests that the composition of complexes is stimulus-dependent. For example, AP-1 binding complexes in the hippocampus (Figure 7.4.) have stimulus-specific compositions that are differentially associated with neuronal death (Kasof *et al.*, 1995). AP-1 proteins are also subject to post-translational modifications; **phosphorylation** can modify both DNA-binding and transactivation functions (Herdegan and Zimmermann, 1995).

Consideration of the complexity of AP-1 function in the brain has encouraged the study of relatively simple systems. One such system (Figure 7.5.) is the rat pineal gland in which trans-synaptic induction of AP-1 gene expression is temporally defined by the action of a single neurotransmitter, noradrenaline, and involves a defined target system, the biosynthesis of melatonin. The hypothalamic oxytocin system has also yielded some important insights into the relationship between c-Fos induction and established neurophysiology (Luckman *et al.*, 1996; see also Guest Box p. 109)

Pathophysiology of Fos and Jun expression in the brain

In addition to the proposed role as mediators of physiological adaptations in neuronal gene expression, AP-1 proteins have also been implicated in pathophysiological mechanisms including epilepsy (Morgan and Curran, 1991), brain injury and drug addiction (Box 7.1)

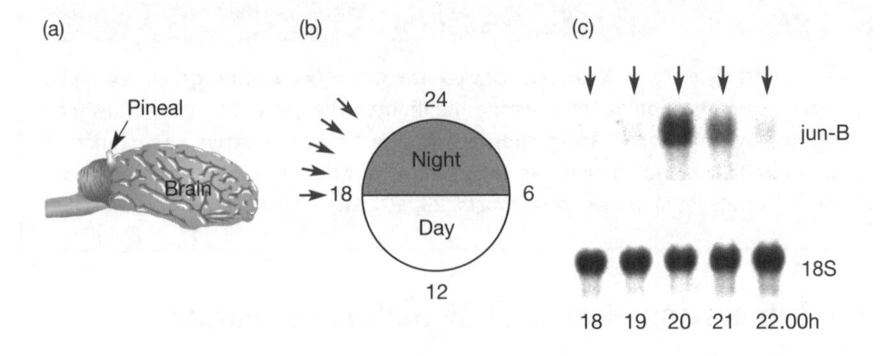

Figure 7.5 Trans-synaptic regulation of gene expression in the pineal gland. (a) Location of the pineal gland. **(b)** Pineal glands are sampled from rats at hourly intervals following lights out at 18.00 h. **(c)** Northern blot showing the transient induction of jun-B mRNA in the pineal gland, peaking at 20.00 h. The levels of 18S ribosomal RNA in the same samples are shown for comparison.

BOX 7.1: MOLECULAR MECHANISMS OF COCAINE ADDICTION

Addiction to psychostimulants such as cocaine and amphetamine remains a major public health problem. Beyond the need to control the illicit supply of such drugs, it is also crucial to research the mechanisms of addiction which engender a craving and associated tendency to relapse that powerfully oppose any attempt at control.

The development of an addiction to drugs such as cocaine appears to involve plasticity in multiple neural circuits (Nestler and Aghajanian, 1997). Cocaine blocks dopamine transporter activity and the behavioural changes associated with addiction are thought to be principally mediated through increased synaptic levels of dopamine in the striatum (but see Caine, 1998). However, the prolonged cellular and molecular changes associated with addiction are not well understood.

Insights into the molecular mechanisms of addiction have been gained through the study of neuronal gene expression in rodents treated with cocaine. Early studies identified changes in the expression of c-*fos* and other IEGs following acute cocaine treatment, and with reference to the proposed role of IEGs (see section 7.3), it was thought that a transient IEG response may underlie a secondary change in functional gene expression. However, more extensive analyses revealed that chronic cocaine treatment was associated with the expression of long-lasting IEG proteins, so-called 'chronic Fos-related antigens (Fras)', later identified as stable isoforms of ΔFosB (Chen *et al.*, 1997). This finding was of great potential relevance to the mechanisms of addiction because Fras were shown to be expressed in groups of striatal neurones over a time course that mirrored cocaine-induced changes in behaviour (Moratalla *et al.*, 1996).

Despite these recent discoveries, much remains to be learnt about the molecular basis of addiction. One way to keep up with the latest advances in addiction research is to check abstracts of the major neuroscience meeting which is held annually in the USA and attracts around 20000 of the world's leading neuroscience

BOX 7.2: Contd

researchers (Society for Neuroscience web site: **http://www.sfn.org/**. For example, a recent presentation at this meeting highlighted the use of inducible transgene technology (Chapter 4) as an approach to question the relative importance of ΔFosB in behavioural changes as compared with other potential cocaine-induced genes (*Society for Neuroscience Abstracts*, **23**, 164.8).

7.4 Integration of nuclear signalling in neurones

Many other nuclear signalling factors, apart from the IEGs discussed above, contribute to the regulation of gene expression in neurones following cellular activation. CREB (see p. 134) is one important factor which is linked to multiple signalling pathways, including the transactivation of c-*fos* (Figure 7.6.). A full consideration of current work in this very active research field is beyond the scope of this book (see Further reading – Segal and Greenberg,

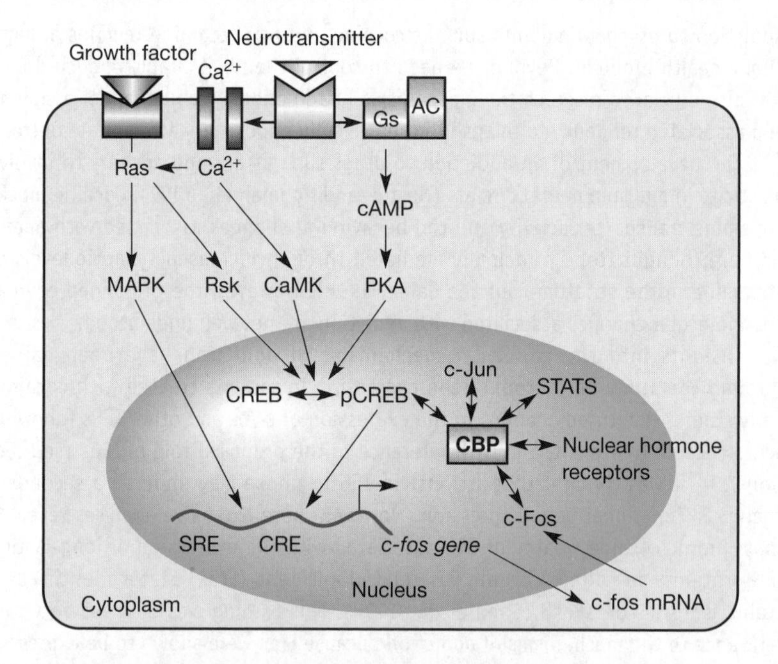

Figure 7.6 Simplified illustration of the cellular signalling mechanisms linked to CREB and transcriptional regulation of the c-*fos* gene. Following stimulation of cell surface receptors by either neurotransmitters or growth factors, second messenger systems are activated, leading to phosphorylation of transcription factors such as CREB which mediate transcriptional activation via binding sites (such as the cAMP response element, CRE) in the c-*fos* promoter. (See Further reading for details of the pathways.) Note that a histone acetyltransferase termed CBP (CREB-binding protein) acts as a co-activator of transcription with many different (only a few examples are shown) transcriptional activators.

1996; Bito *et al.*, 1997; Goodman and Mandel, 1998). With respect to Fos, it is also informative to refer to the Home page of Michael Greenberg, a pioneer in this field who has continued to make important contributions to the understanding of neuronal signalling pathways: **http://www.fas.harvard.edu/~ashaywit/index.html**.

7.5 Summary

- Differentiated neurones exhibit changes in gene expression.
- Expression of the vasopressin gene in the hypothalamus illustrates many of the characteristics of neuronal gene plasticity, including changes in mRNA level *and* size, transcriptional and post-transcriptional modes of regulation, and cell-specific responses.
- Neurotransmitter receptor genes also exhibit physiological changes in expression.
- Cellular immediate-early genes are induced in the brain following a variety of stimuli, and may form a link between cellular activation and functional genomic responses.
- Fos, Jun and other members of the AP-1 family are implicated in many neuronal mechanisms including cell death, neurotrophic factor signalling, pain mediation, drug addiction and learning and memory.
- The nuclear signalling mechanisms which control changes in neuronal gene expression are highly integrated, and partially characterised.

Further reading

BITO, H., DEISSEROTH, K. and TSIEN, R.W. (1997) Ca^{2+}-dependent regulation in neuronal gene expression. *Current Opinion in Neurobiology* 7, 419–429.

SEGAL, R.A. and GREENBERG, M.E. (1996) Intracellular signalling pathways activated by neurotrophic factors. *Annual Review of Neuroscience*, **19**, 463–489.

Recent reviews by prominent research groups working on neuronal signalling mechanisms.

GOODMAN, R.H. and MANDEL, G. (1998) Activation and repression in the nervous system. *Current Opinion in Neurobiology* **8**, 413–417.

Recent studies of transcriptional co-activators such as CBP, and of transcription repressors.

NESTLER, E.J. and AGHAJANIAN, G.K. (1997) Molecular and cellular basis of addiction. *Science*, **278**, 58–63

A brief review of recent work on the molecular and cellular basis of drug addiction.

WEN, X., FUHRMAN, S., MICHAELS, G.S., CARR, D.B., SMITH, S., BARKER, J.L., SOMOGYI, R. (1998) Large-scale temporal gene expression mapping of central nervous system development. *Proceedings of the National Academy of Sciences, USA*, **95**, 334–339.

Describes the use of RT–PCR technology (Box.6.4) to measure temporal changes in gene expression, identifying 'waves' of expression by cluster analysis.

CHAPTER EIGHT

The molecular basis of behaviour

Key topics

- Learning and memory
- Memory tests in rodents
- The hippocampus
- Long-term potentiation
- Candidate learning and memory genes
  - Neurotransmitters
  - Cellular signalling molecules
  - Retrograde messengers
- Knockout memory models
- Inducible knockout models
- Circadian rhythms
- Forward genetic strategy
- Transgenic rescue
- Transcriptional feedback model of circadian timing
- Antisense transcript model of circadian timing

8.1 Introduction

Molecular neuroscientists are now addressing the basic mechanisms of behaviour across a broad front – important aspects of this major research effort will be illustrated here with reference to two fields:

- Learning and memory
- Circadian rhythms

Before attempting to understand specific aspects of behavioural mechanisms it is necessary to ask – what do we mean by 'the molecular basis of behaviour'?

- The gene that is required for the expression of a particular behaviour.
- The biological action and regulation of the protein product of the gene.
- The assembly of multiple gene products into an ordered arrangement of mechanisms that is sufficient for the manifestation of a particular behaviour.

Any of these answers is acceptable but, although each may form the basis of an experimental hypothesis, the first two are of limited application for an understanding of the mechanistic basis of mammalian behaviours. Thus, it is clear that the neural mechanisms which underlie behaviours are multicomponent, or **polygenic** with respect to the genes which are involved. The behaviours discussed in this chapter have been selected to demonstrate current experimental approaches towards a molecular basis, but also illustrate the difficulty of defining this end-point. On the subject of end-points, the following story (Geertz, 1973) is pertinent:

> There is an Indian story – at least I heard it as an Indian story – about an Englishman who, having been told that the world rested on a platform which rested on the back of an elephant which rested in turn on the back of a turtle, asked....what did the turtle rest on? 'Another turtle.' And that turtle? 'Ah, Sahib, after that it's turtles all the way down.'

8.2 Learning and memory

Cognition is the act or process of knowing that involves the processing of sensory information and includes perception, awareness, judgement, learning and memory. Of all the aspects of cognition, the mental process that really defines us as individuals is our ability to recall past events – our ability to remember people, places and ideas, and to recall those things at will – in other words, learning and memory.

Learning is the acquisition of knowledge or skill, or the modification of a behavioural tendency by experience. **Memory** is what has been learnt; mechanistically, the storage and recall of things learned and retained from the experience of an organism.

Cognitive psychological studies suggest that, in humans, memory consists of at least two distinct mental processes – implicit and explicit learning:

- **Implicit learning** refers to motor skills and perceptual strategies, for example, learning how to swim, or how to ride a bicycle. Anatomically, implicit learning is localised to the sensory and motor systems, and is thought to be related to the habituation or sensitisation of reflex pathways.
- **Explicit learning** refers to our memories of people, places and things.

Can the experience of human explicit learning be related to forms of cognition in animal models that can be studied mechanistically and quantitatively? Explicit learning is thought to be related to a complex form of associative memory called configural memory.

Associative memory involves learning to associate a stimulus with a response or event. Two types of associative learning have been defined in experimental models – procedural or simple memory and declarative or configural memory:

- **Simple memory** involves the association of a single cue with a response.
- **Configural memory** is complex, and involves the configuration or integration of multiple cues or facts.

8.2.1 Memory tests in rodent models

Numerous behavioural tests have been developed that enable simple memory and configural memory to be distinguished and quantitated in mice and rats. One of the most effective and exploited tests is the Morris watermaze navigation test (Box 8.1).

8.2.2 The anatomy of memory

Patients suffering brain lesions caused by, for example, stroke, surgery, viral infection, acts of violence or accidents, have been used to map the structures in the human brain required for memory formation. The brains of patients have been examined upon death, and any region lost or damaged can be retrospectively correlated with psychological deficits. Patients with hippocampal lesions lose the capacity to form new long-term memories, but old memories are retained and can be recalled. Short-term memory is intact, as is the ability to develop new motor skills. Thus, the hippocampus appears not to be involved in implicit learning, short-term memory, memory storage or memory recall. Rather, the hippocampus is required for the formation of new, long-term explicit memories.

Similar studies have been performed in rodent models. Using the **Morris watermaze** (Box 8.1), rats with hippocampal lesions were shown to have lost their capacity for spatial learning, while their non-spatial learning was intact. In memory tests, presentation of the stimulus after varying lengths of time after training can test short- or long-term memory. Thus memory has been classified into:

- immediate (up to 10 minutes)
- short-term (up to 1 hour, and not dependent upon synthesis of new RNA or proteins)
- long-term (lasting a life-time, and dependent upon new gene expression)

BOX 8.1: THE MORRIS WATERMAZE

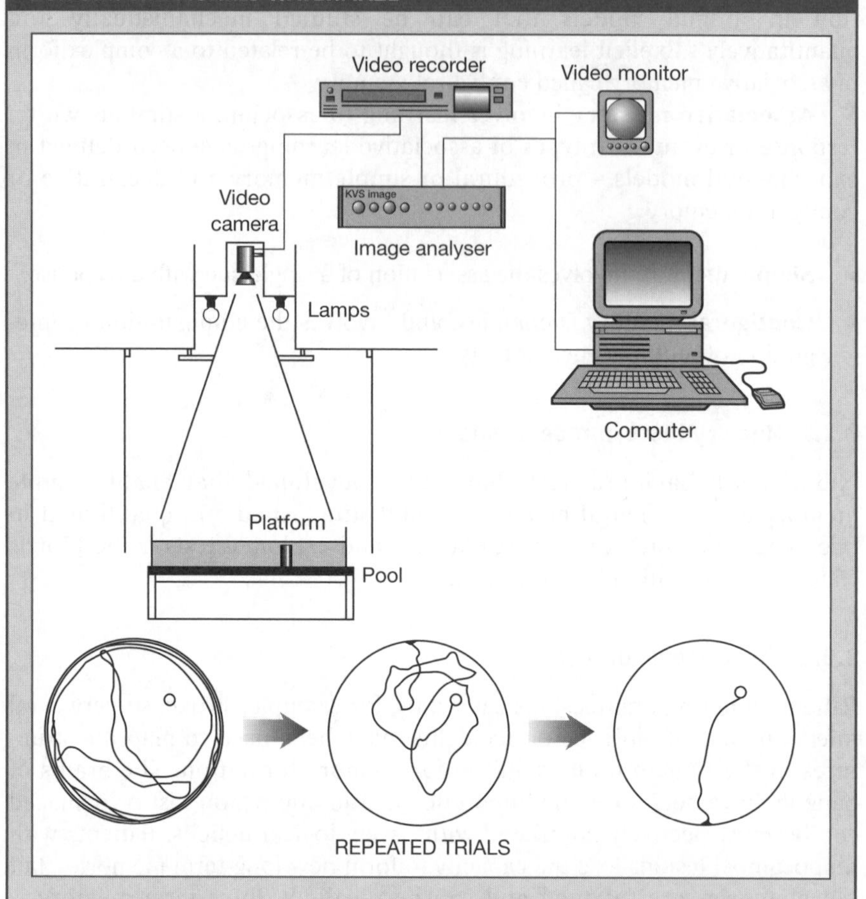

The Morris watermaze **(http://sirius.lns.ed.ac.uk/community/members/morris.html; http://www.watermaze.com)** assesses an animal's capacity for either spatial memory, a form of configural learning, or non-spatial memory, a form of simple learning. Spatial memory involves the formation of an internal representation of the environment, whereas non-spatial memory involves the association of a simple visual cue with a response. The Morris watermaze consists of a circular pool filled with opaque water. The escape route consists of a platform submerged 1 cm below the water surface. The subject is placed in the pool at one of four sites, and must learn to navigate to the platform. Two variations of the test have been developed which are similar in terms of motivation, visual ability and motor skills, but which distinguish between and quantitatively assess spatial versus non-spatial learning. In the hidden platform test, the platform is kept hidden below the surface of the water. Animals are placed in the pool at one of four sites, but the position of the platform is kept constant between trials. To escape, the animals must learn to navigate to the hidden platform by mapping its position relative to visible cues outside of the pool. Initially, the animal will take time to explore the pool and chance upon the platform. However, in subsequent trials, an animal with

BOX 8.1: Contd

an intact capacity for spatial memory will associate the position of the platform with the visible cues, and will escape from the water with increasing rapidity and ease. In the visible platform test, a clearly visible landmark, such as a flag, is placed on the platform to indicate its position. Animals are placed in the pool at one of four sites, but the platform is moved between trials. The subjects must therefore learn to associate the landmark with the platform; spatial cues are irrelevant. However, the visual ability, motivation and motor skills required for the visible platform test are equivalent to those of the hidden platform test.

The hippocampus is not involved in short-term memory, nor is it involved in long-term memory storage or recall, although there is some evidence that memories are held in the hippocampus for up to four weeks prior to permanent storage elsewhere. Rather, the hippocampus appears to be involved in the conversion of short-term memories into long-term memories.

The circuitry of the hippocampus is composed of three major connected pathways (Figure 8.1). Cortical input into the hippocampus is via the perforant pathway from the entorhinal cortex (Figure 8.1) to a structure in the hippocampus called the dentate gyrus. The granule cells of the dentate gyrus project to the large pyramidal cells of the CA3 region via the

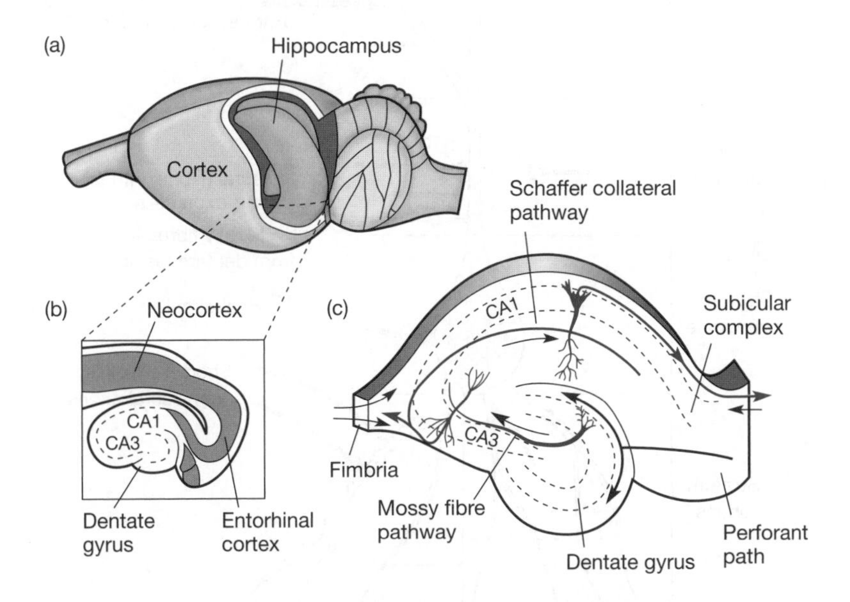

Figure 8.1 (a) Location of the hippocampus within the rat brain. (b) The hippocampus, when cut transverse to its longitudinal axis, exhibits a strong afferent set of three connected pathways. (c) Layers II and III of the entorhinal cortex project to the granule cells of the dentate gyrus, via the perforant path. The granule cells of the dentate gyrus project to the large pyramidal cells of Ammon's horn, subfield 3 (CA3), via the mossy fibre system. CA3 pyramidal cells project to the pyramidal cells of the CA1 subfield, via the Schaffer collateral system. (Continued overleaf.)

mossy fibre system. Finally, the CA3 pyramidal cells project to the pyramidal cells of the CA1 subfield, via the Schaffer collateral system. Damage to any one of these pathways can disrupt memory formation. Within the CA1, different afferent pathways make synaptic contact with pyramidal cells at different locations (Figure 8.1d).

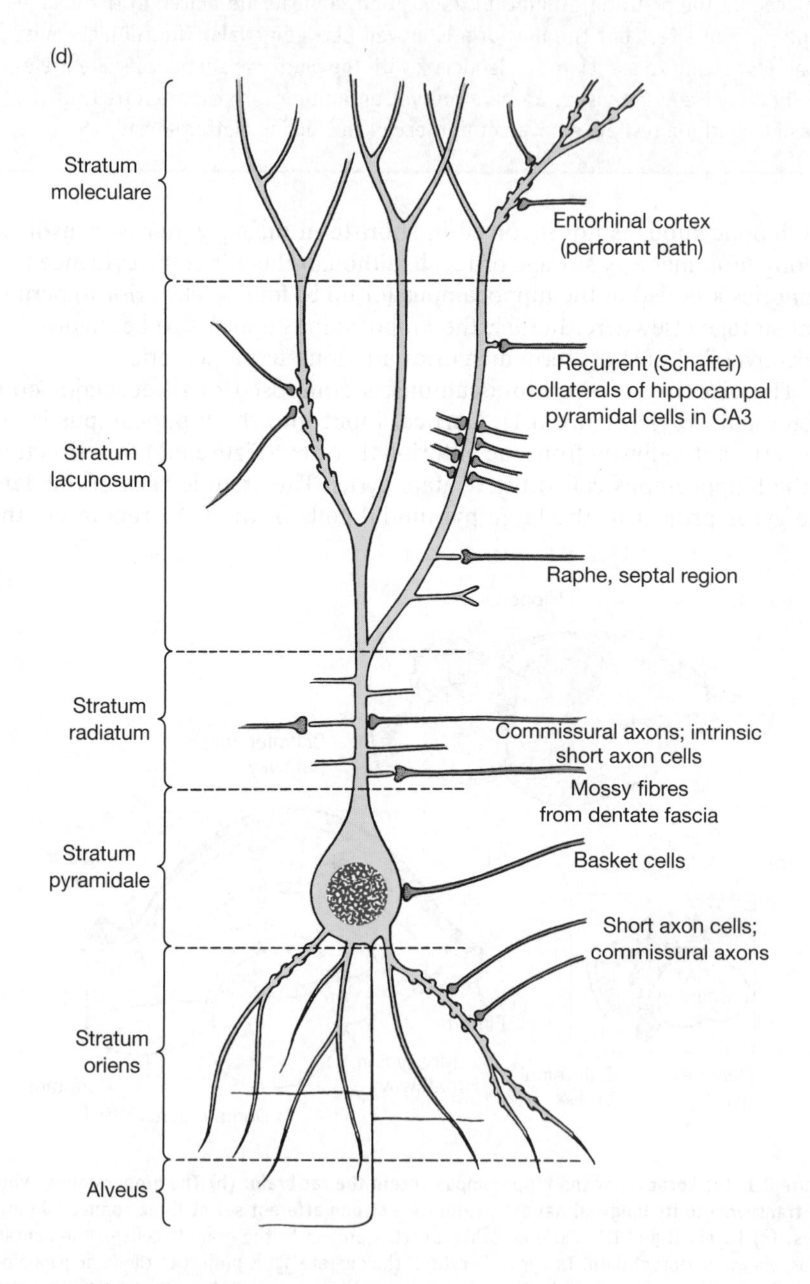

Figure 8.1 (d) Structure of a CA1 pyramidal cell showing that different afferent pathways make synaptic contact at different locations.

8.2.3 Long-term potentiation

In 1949, Donald Hebb (Hebb, 1949) suggested that the physiological basis of memory involved the long-term strengthening of synapses between neurones that are coincidentally active. Subsequently, a form of synaptic plasticity called long-term potentiation (LTP) was discovered (see Bliss and Collingridge, 1993). Basically, LTP means that the efficiency of a synapse increases if that synapse is active.

LTP is present in the hippocampus and, because of the association between this structure and memory, it has been suggested that LTP may represent a mechanism for memory formation.

LTP can be measured *in vivo,* but such experiments are difficult. More convenient is the use of cultured brain slices. A stimulatory electrode is placed in, say, the CA3 pyramidal cells, and used to deliver an excitatory pulse which is transmitted via the Schaffer collateral pathway to the CA1 region. A recording electrode measures the postsynaptic current (PSC) in CA1 neurones. After an excitatory pulse, a second pulse is delivered and, again, the current is measured – any increase in the postsynaptic current is LTP.

Slice cultures can readily be dosed with drugs that can be used to assess the role of the target molecules in LTP. Thus, LTP can be divided into a number of temporal categories based on their sensitivity to drugs affecting various cellular functions:

- Short-Term Potentiation (STP) lasts for 30–60 minutes, and is independent of inhibitors that block the activity of cellular protein kinases.
- LTP1 (or early LTP; E-LTP) lasts for 3–6 hours, and is blocked by kinase inhibitors.
- LTP2 is dependent upon translation – the synthesis of new proteins.
- LTP3 (or late LTP; L-LTP) lasts for many days, and is dependent upon transcription – the synthesis of new mRNAs.

These observations are strikingly reminiscent of the time course of immediate, short- and long-term memory.

A model for LTP

Following the arrival of an action potential at the presynaptic terminal, neurotransmitter is released in discrete packets, or quanta (Figure 8.2). A synaptic connection might involve one or several sites that can release quanta. These sites function in a probabalistic fashion; upon the arrival of an action potential, each site releases a quantum with a certain probability known as the release probability. The neurotransmitter in each packet then acts on receptors at the postsynaptic terminal and produces an electrical response with a characteristic (quantal) size.

Considering this model, LTP might be a consequence of events at the pre- or postsynaptic terminal, or both. At the presynaptic terminal, LTP could be due to an increase in:

- the number of release sites (Figure 8.2, stage 1); or,
- release probability (Figure 8.2, stage 2),

both of which would increase the average number of quanta released per action potential, and are properties of the presynaptic terminal. Such a model must invoke the action of a retrograde message from the postsynaptic terminal to the presynaptic terminal. Alternatively, LTP might be due to:

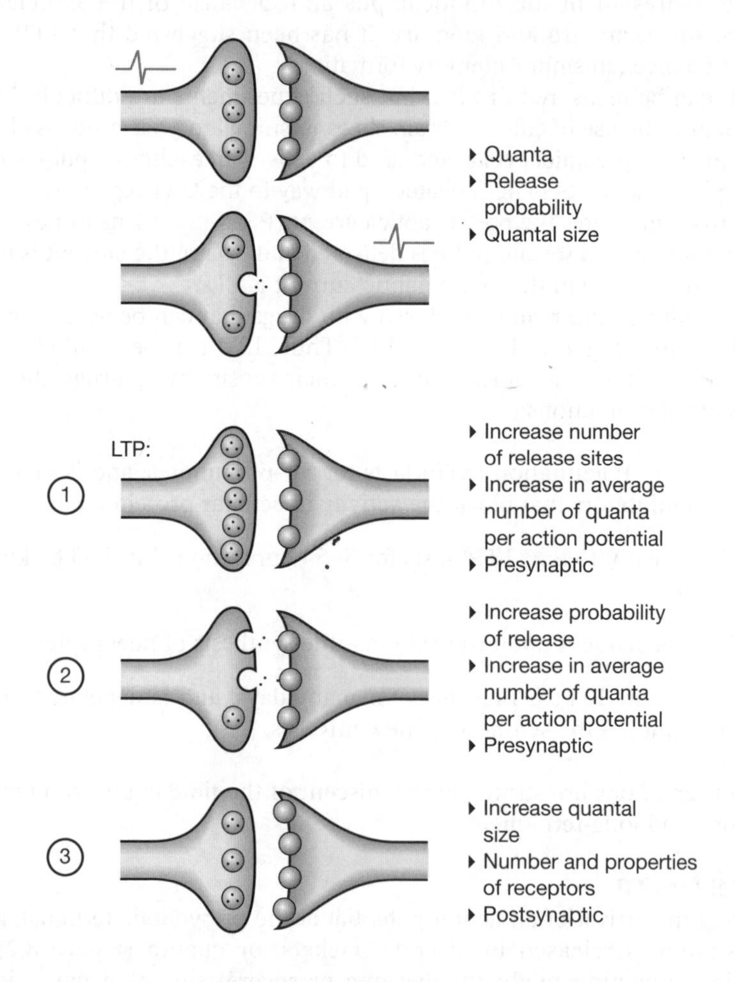

Figure 8.2 A model for LTP. Following the arrival of an action potential at the presynaptic terminal, neurotransmitter is released in discrete packets, or quanta. A synaptic connection might involve one or several sites that can release quanta. These sites function in a probabilistic fashion; upon the arrival of an action potential, each site releases a quantum with a certain probability known as the release probability. The neurotransmitter in each packet then acts on receptors at the postsynaptic terminal and produces an electrical response with a characteristic (quantal) size. LTP might thus be a consequence of:

1. An increase in the number of release sites.
2. An increase in release probability.
3. An increase in the quantal size.

- an increase in the quantal size (Figure 8.2, stage 3),

which is determined by number and properties of receptors on the post-synaptic terminal.

8.2.4 Molecules and memory – a general strategy

How are the latest molecular techniques being used to study LTP, and elucidate its role, if any, in learning and memory? An integrated approach has been adopted. Numerous candidate learning and memory genes have been identified on the basis of:

- biochemical studies – are the functional properties of a protein consistent with a role in information storage?
- expression studies – is the gene expressed in the hippocampus, and does it have a subcellular localisation consistent with a role in LTP?
- pharmacological studies – is LTP affected when the activity of the protein is altered by drugs?

In what other ways could learning and memory genes be identified?

Such studies have identified some of the synaptic processes involved in hippocampal LTP, and have suggested the involvement of candidate memory and learning genes. It has been difficult to deduce the contribution of these proteins to LTP for three reasons:

1. The drugs are rather non-specific, and have effects on enzymes or processes other than those that were intended.
2. The targeted proteins are involved in a multitude of cellular processes which are unrelated to synaptic function and memory.
3. The targeting of drugs to particular cell types or subcellular structures is not possible.

To overcome these difficulties, and to introduce a new level of specificity into learning studies, modifications – usually null mutations – of these genes have been introduced into mice using gene transfer or **knockout** technologies (see Chapter 4), and the effect on hippocampal development, LTP, learning and memory has been assessed. For the first time it has been possible to relate specific biochemical defects to deficits in LTP which in turn can be related to effects on learning and memory processes in the whole animal. Importantly, problems with the germline approach, detailed in Chapters 4 and 5, have been overcome by the use of technologies that enable cell-type and inducible genetic lesions to be introduced into the whole animal. Much of this exciting work has been performed by a single group – that of Eric Kandel at Columbia University, New York: **(http://cpmcnet.columbia.edu/dept/neurobeh/Kandel.html).** In addition, mice with engineered cognitive deficits are being used in biochemical studies to elucidate the pathways that interact with the modified gene product in the mediation of LTP (see Guest Box p. 126).

GUEST BOX BY SETH G.N. GRANT

Understanding the molecular mechanisms of synaptic transmission

Understanding the molecular mechanisms of synaptic transmission is essential for a comprehensive picture of how neuronal networks generate behaviour. A central experimental paradigm for examination of the links between molecules, synapses, cells and systems neuroscience is found in the study of learning and memory in the mouse. Currently, the mouse is the only organism where it is possible to perform electrophysiological and biochemical experiments on genetically manipulated specific neurones involved in learning.

A large body of literature supports the view that 'activity-dependent' changes in synaptic transmission, known as synaptic plasticity, are an essential component of learning. Synaptic plasticity is readily measured in mouse hippocampus and high-frequency trains of action potentials can induce long-lasting synaptic potentiation (long-term potentiation, LTP) or depression (long-term depression, LTD). In the hippocampus, as in many other regions of the brain, LTP and LTD require the activation of the postsynaptic NMDA subtype of glutamate receptor. This receptor can activate a variety of cellular responses, including LTP and LTD which indicates that it couples to multiple complex signal transduction pathways. A basic challenge for the experimentalist is to identify the molecules involved in synaptic plasticity and explain how they form molecular signalling cascades. This is where genetics makes a powerful contribution, since the combination of genetics and biochemistry with physiology can define pathways.

To explore the role of tyrosine kinases in synaptic plasticity we examined mice carrying mutations in four kinase genes (*Fyn*, *Src*, *Yes*, *Abl*) that are expressed in the brain as well as pharmacological antagonists of tyrosine kinases. Both the antagonists and *Fyn* mutant mice showed impairments in the induction of LTP. A simple way to identify possible downstream components of a Fyn signalling pathway is to examine hypophosphorylated proteins, and we found that Focal Adhesion Kinase, the NMDA receptor subunits and NMDA receptor associated proteins are underphosphorylated. This provides direct evidence that Fyn affects NMDA receptor signalling, perhaps by modulation of associated proteins.

We tested if NMDA receptor-associated proteins are involved with synaptic plasticity by examining mice carrying mutations in Post Synaptic Density 95 protein, which binds to NMDA receptor cytoplasmic domains. These mice show a striking increase in LTP and a robust LTP at frequencies of stimulation that normally elicit LTD. This suggests that PSD95 couples a negative regulator to the NMDA receptor signalling pathway.

As more proteins are found in these signalling pathways and their corresponding mutants created, we will be able to take classical genetic approaches using

GUEST BOX Contd

double-mutant mice to complement the biochemical studies to define the biochemical pathway. Finding these pathways at the synapse will no doubt lead us into exciting new insights into one of the most exquisitely regulated forms of cell–cell communication and open new ways to understand cognition.

GRANT, S.G.N., LARL, K.A., KIEBLER, M.A. and KANDEL, E.R. (1995) Focal adhesion kinase in the brain: novel subcellular localization and specific regulation by fyn tyrosine kinase in mutant mice. *Genes and Development*, **9**, 1909–1921.

Dr Seth G.N. Grant is at the Centre for Genome Research, University of Edinburgh, and is Director of the Centre for Neuroscience. Dr Grant received his medical training at the University of Sydney and did postdoctoral research at Cold Spring Harbor Laboratory, USA, with Dr Douglas Hanahan working on transgenic mice and oncogenes. He then worked with Dr Eric Kandel at Columbia University, New York City, where he used mutant mice to study the molecular mechanisms of learning and memory.

8.2.5 Memory and molecules – synaptic transmission

LTP in the CA1 region of the hippocampus has been shown to be triggered postsynaptically by an increase in intracellular Ca^{2+}. This is brought about by *N*-methyl-D-aspartate (NMDA)-type glutamate receptors (NMDARs) – transmitter-gated ion channels that interact with L-glutamate released from the presynaptic terminal **(http://www.euro.promega.com/nnotes/nnotedex.html#ii3; http://bioinfo.weizmann.ac.il/hotmolecbase/entries/nmda.htm**. This results in a Ca^{2+} influx through the opened receptor (Figure 8.3). The NMDARs consist of a complex family of heterodimers consisting of one of the eight NMDAR1 subunit splice variants (a–h), and one of the four NMDAR2 isoforms (A–D). Gene knockout experiments have confirmed pharmacological studies that had identified the NMDARs as crucial mediators of LTP. Furthermore, these experiments have directly correlated an LTP deficit with reduced performance in spatial memory tests. Mice lacking the NMDA2a subunit survive and develop normally, but show reduced NMDA synaptic currents, reduced LTP and impaired spatial learning as assessed by performance in the Morris watermaze. Mice lacking a functional NMDAR1 gene die within a day of birth, thus precluding studies on LTP and memory. This problem has been overcome by the use of technologies that confine the knockout to specific parts of the brain (see Box 8.2).

Can you think of other ways to study the role of the NMDAR1 gene in the adult hippocampus?

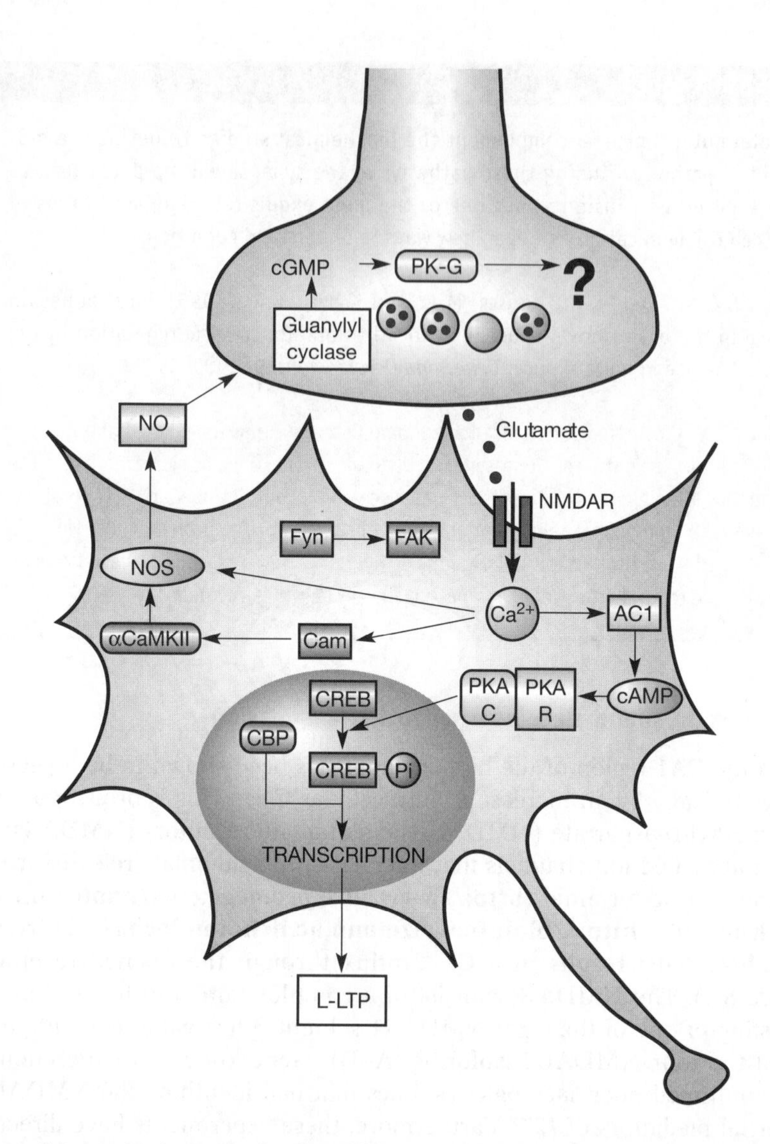

Figure 8.3 A molecular mechanism for LTP? Known and inferred pre- and postsynaptic molecular processes thought to be involved in LTP. NMDAR, NMDA receptor; AC1, adenylate cyclase isoform 1; PKA R, regulatory subunit of protein kinase A; PKA C, catalytic subunit of protein kinase A; CREB, cAMP-response element binding protein; CBP, CREB binding protein; L-LTP, late phase LTP; αCamKII, α Calcium-calmodulin-dependent kinase II; Cam, calmodulin; NOS, nitric oxide synthase; NO, nitric oxide; PK-G, cGMP-dependent protein kinase.

8.2.6 Molecules and memory – intracellular signalling

Downstream of Ca^{2+}, a variety of postsynaptic intracellular signalling molecules have been identified as being encoded by candidate learning and memory genes (Figure 8.3) by virtue of the effect of kinase inhibitors on LTP.

BOX 8.2: CELL-SPECIFIC KNOCKOUT OF THE NMDAR1 GENE

A transgenic mouse line was derived in which Cre recombinase was expresse pyramidal cells of the hippocampal CA1 region under the control of an 8.5 kb p moter from the α-Calcium-Calmodulin-dependent Kinase II (αCamKII) ge which directs expression to neurones in specific brain regions, including the hi pocampus (see Figure 4.4). Using ES cell-mediated gene targeting technologies (see Figure 4.4 – just substitute the NMDAR1 gene for Gene X), a separate line of mice was generated in which *loxP* sequences were introduced into the NMDAR1 gene (see section 4.5.1). Bitransgenic mice were then derived in which the Cre protein mediated a large deletion of a crucial region of the NMDAR1 gene, but only in cells of the CA1 region. This time the mutant mice grew to adulthood without revealing any obvious abnormalities. These mice lacked NMDA receptor-mediated synaptic currents and LTP in CA1 synapses, corresponding to the cellular location of the NMDAR1 knockout, and exhibited impaired spatial, but not non-spatial, learning (Tsien *et al.*, 1996).

αCamKII

αCalcium-calmodulin-dependent kinase II (αCamkII) is a serine/threonine kinase that is activated through autophosphorylation on Thr^{286} by Ca^{2+}-loaded calmodulin (Cam). The activity of autophosphorylated αCamKII is no longer Cam- or Ca^{2+}-dependent, and the enzyme remains in an active state well beyond the duration of the activating Ca^{2+} signal. Evidence from three lines of study led to the suggestion that αCamKII may be a molecular substrate for memory:

- **Anatomy** – αCamKII is neurone-specific, and is found in abundance adjacent to postsynaptic membranes in synapses that exhibit LTP.
- **Pharmacology** – inhibitors of αCamKII block the formation of LTP.
- **Biochemistry** – a transient Ca^{2+} flux elicits a persistent αCamKII response.

In order to test the hypothesis that αCamKII is involved in LTP, learning and memory, the gene encoding this enzyme has been manipulated in two ways in mice. Firstly, mice devoid of αCamKII have been made using germline knockout technology (see Chapter 4). Secondly, a constitutively active (Ca^{2+}-independent) mutant form of αCamKII has been expressed in mice in both a cell-specific and an inducible fashion.

αCamKII knockout

αCamKII-null mice are viable, and the mutation has no apparent effect on the development or gross anatomy of the brain. While the mice have normal postsynaptic mechanisms, LTP is considerably reduced, although it is not eliminated altogether. Most synapses had no measurable LTP, but those few synapses that did exhibit LTP gave normal responses. These data suggest that the absence of αCamKII might increase the threshold of a rate-limiting process in the induction of LTP. The performance of the mutant mice was then assessed in the Morris watermaze; the mice showed normal non-spatial learning, but impaired spatial learning. Interestingly, performance in the hidden platform test improved with repeated training, suggesting a threshold effect that is reminiscent of the LTP studies.

Expression of mutant αCamKII

Are the phenotypes described in the αCamKII knockout mice a consequence of a specific effect on LTP, learning and memory processes within the hippocampus, or are they the result of anatomically or developmentally distant effects? In an attempt to obviate the problems inherent in germline transgenesis and knockout technology, a mutant form of αCamKII has been expressed in the mouse brain, firstly in a cell-specific fashion, and latterly in a cell-specific and inducible fashion.

A mutation of Thr^{286} to Asp converts αCamKII to a constitutively active, Ca^{2+} -independent form. αCamKII regulatory sequences were then used to direct expression of the mutated kinase to specific parts of the mouse brain, particularly in the hippocampus (Figure 8.4). No mutant kinase was expressed in other parts of the brain (Figure 8.4). LTP was found

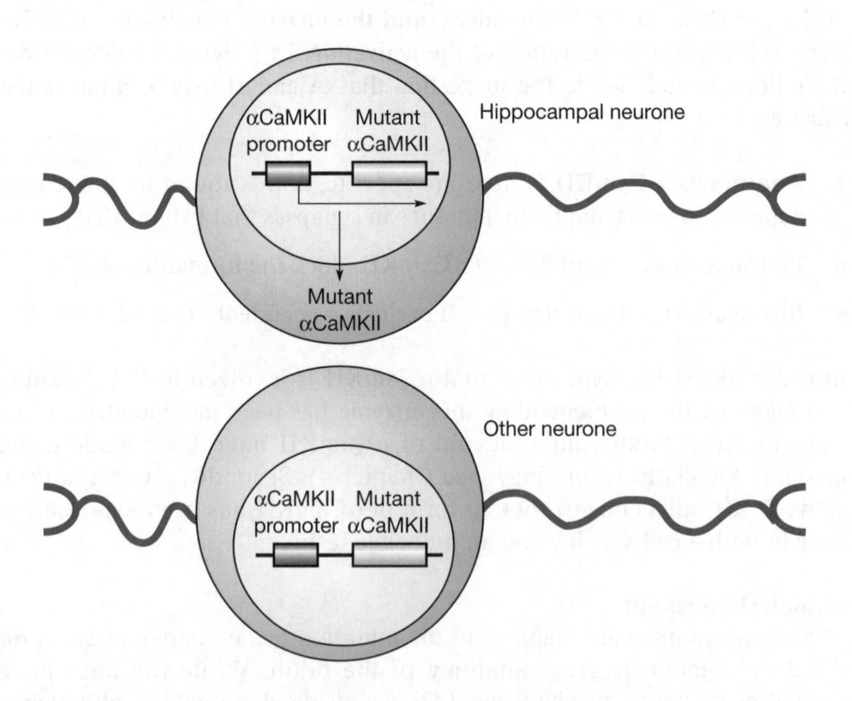

Figure 8.4 Hippocampal neurone-specific expression of a Ca^{2+}-independent constitutively active form of αCamKII. The αCamKII promoter directs expression of the mutant to hippocampal neurones, but not to other types of neurone.

to be eliminated at a frequency (10 Hz) found in the hippocampus of animals during spatial exploration. This deficit in LTP correlated with a defect in spatial memory. These are cell-specific effects – the mutant αCamKII is expressed in only a restricted number of cells in the mouse brain. However, are these a direct effect on memory and its underlying synaptic mechanism, or an indirect effect on the development of the neuronal circuitry? To address this issue, it would be useful to be able to turn the mutated αCamKII transgene on and off at will and look at the phenotypic effects. This is precisely what the Kandel group have done using the tetracycline gene regulation system (Box 8.3).

Non-receptor tyrosine kinases

The family of non-receptor tyrosine kinases comprises at least nine members, of which four, Src, Fyn, Yes and Abl, are present in synaptic plasma membranes and in the postsynaptic density. Inhibitors of tyrosine kinases block the induction of LTP. However, these drugs lack the pharmacological specificity needed to distinguish between different members of either the receptor or non-receptor families. In an attempt to identify the specific kinases that are involved in LTP, and perhaps learning and memory, mice with null mutations in the *src, fyn, yes* and *abl* genes have been examined. Only the Fyn knockout mice showed cognitive deficits; LTP was blunted and spatial learning was lost. Interestingly, the *fyn* mutant mice showed a graded reduction in their ability to induce LTP. Using a low-amplitude tetanic stimulation, no LTP was seen, whereas a high-amplitude tetanic

BOX 8.3: CELL-SPECIFIC AND INDUCIBLE EXPRESSION OF AN αCAMKII MUTANT

A transgenic mouse line was derived in which **Tet-OFF** (see Chapter 4) was expressed in specific brain regions, particularly the hippocampus, under the control of an 8.5 kb promoter from the αCamKII gene. A second mouse line contained a transgene consisting of the Thr^{286} to Asp mutant αCamkKII cDNA under the control of a Tet-responsive promoter. Bitransgenic (see Figure overleaf) mice were then derived in which the mutant αCamKII was expressed, but only in those neurones that express **Tet-OFF**, and then only in the absence of **Dox**. In the rest of the brain, where **Tet-OFF** was not expressed, there was no expression of the mutant αCamKII transgene. Of course, the mice develop, and are born, without exposure to Dox; they thus develop with the expression of the mutant αCamKII in the hippocampus and, as a consequence suffer from LTP and learning deficits. There is a loss of LTP in the 10 Hz range, and a deficit in spatial memory. However, upon suppression of transgene expression by the application of Dox, the mice recovered LTP and spatial memory; these were lost again following Dox withdrawal. These data strongly suggest a role for hippocampal αCamKII in memory formation, independent of any developmental function (Mayford *et al.*, 1996)

BOX 8.3: Contd

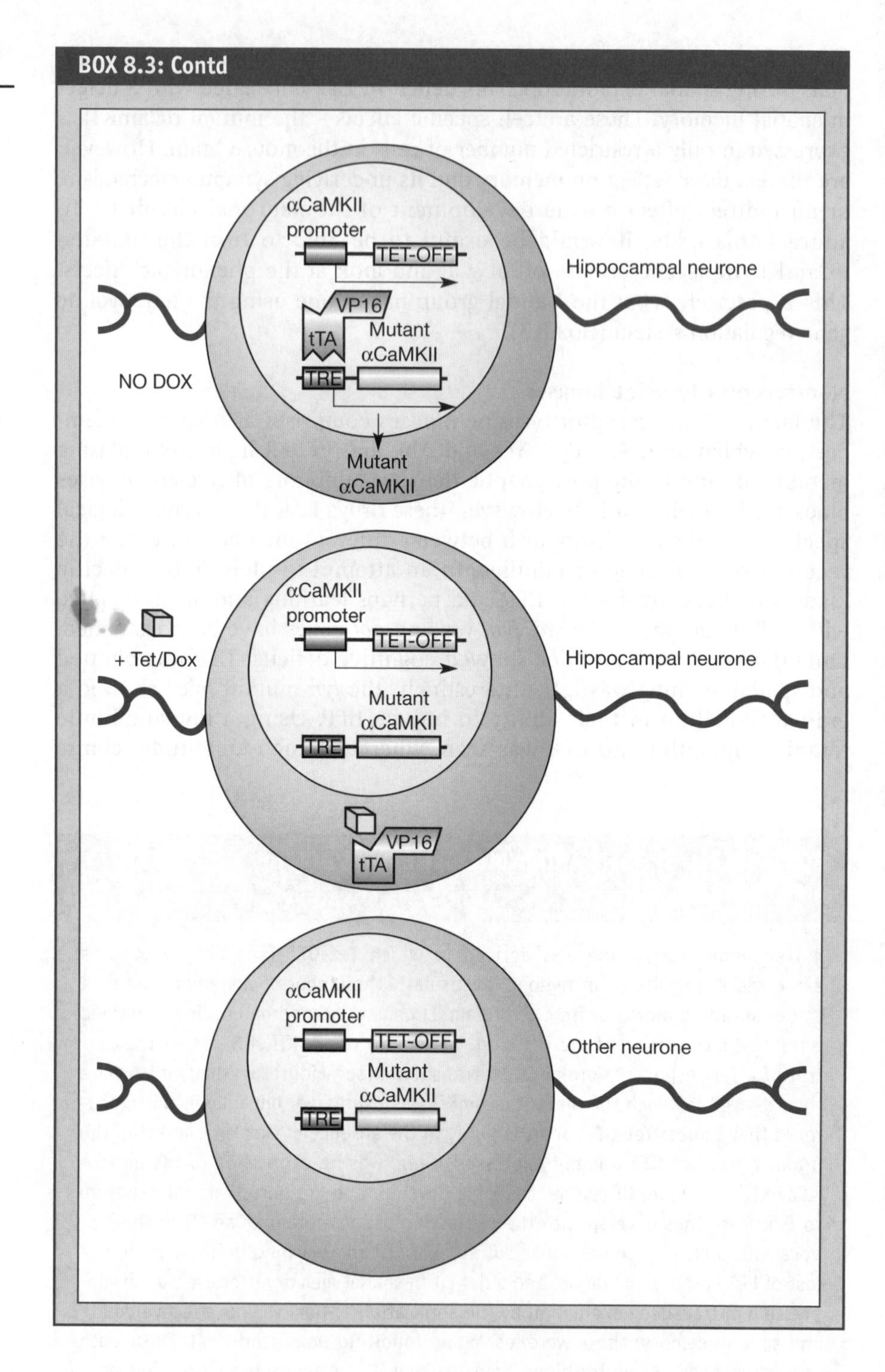
αCaMKII
promoter
TET-OFF
Hippocampal neurone
VP16
tA
Mutant
αCaMKII
NO DOX
TRE
Mutant
αCaMKII
αCaMKII
promoter
TET-OFF
+ Tet/Dox
Hippocampal neurone
Mutant
αCaMKII
TRE
VP16
tTA
αCaMKII
promoter
TET-OFF
Other neurone
Mutant
αCaMKII
TRE

stimulation was able to induce LTP, albeit at a reduced level. This observation led to the suggestion that a compensatory mechanism, perhaps another kinase, was recruited following a high-intensity stimulus. This was the first indication that another family member could substitute for one that had been knocked-out; a concept supported by the observation that the LTP induced by high-intensity stimulation in *fyn* knockout mice could be inhibited by tyrosine kinase inhibitors.

One complication of the *fyn* knockout model is that this gene deletion resulted in a developmental defect that complicated interpretation of the synaptic plasticity data. While the gross architecture of the hippocampus was normal, defects in the arrangement of the granule layer of the dentate gyrus, and of their target cells in the CA3 region, were observed in the knockout mice. It was thus possible that the impairment of LTP in Fyn-deficient mice was a consequence of impaired neuronal development, rather than deficits in synpatic plasticity in mature hippocampal neurones. In order to distinguish between these two possibilities, Fyn-rescue mice were generated by the expression of a transgene encoding wild-type Fyn in hippocampal neurones of Fyn-deficient mice. Although the morphological abnormalities were still evident in the Fyn-rescue mice, LTP was restored, suggesting that hippocampal Fyn contributes directly to the molecular mechanism of LTP induction (Kojima *et al.*, 1997).

What are the molecular defects in *fyn* mutant mice?

The pattern of tyrosine phosphorylation of brain proteins was compared in *src, yes* and *fyn* mutant mice, and nine proteins were found to be hypophosphorylated in Fyn null animals. One of these proteins was identified as focal adhesion kinase (FAK; Grant *et al.*, 1995), which was originally identified in focal cell adhesions of Src transformed fibroblasts. In *fyn* mutant mice, FAK protein levels are unchanged, but FAK is less phosphorylated. FAK autophosphorylation increases its affinity for SH2 domains of the Src family of tyrosine kinases, which then phosphorylate FAK, and this correlates with increased FAK kinase activity. In the absence of Fyn, reduced phosphorylation of FAK leads to reduced FAK activity.

Is FAK a downstream component of the LTP and learning pathways mediated by Fyn?

The FAK gene has been knocked out in mice, but gross mid-gestation defects, including a general failure of mesoderm development, preclude studies on adult cognitive function. FAK is thus an obvious target for cell-specific or inducible gene manipulation.

What other experiments could be performed to examine the role of FAK in the adult hippocampus?

Why do only the *fyn* knockouts exhibit a synaptic phenotype?

FAK was originally identified as a Src substrate and it associates with, and is phosphorylated by, all of the members of the Src family. If FAK is a downstream mediator of Src family tyrosine kinase activity in LTP and memory, why is it that only the *fyn* null mutation elicits an effect on these processes? The answer is that different family members are able to compensate for each other to a greater or lesser degree. Thus, Fyn and Yes activity is increased in Src mutants, and may functionally compensate for the Src deficit. Src activity is increased in Fyn mutants, but cannot compensate sufficiently for the deficit. Mice doubly mutant for Fyn and Src exhibit a more drastic phenotype than either of the single mutants (although the mice die perinatally, cognitive function has not been examined). However, Fyn/Yes double-mutant mice have the same LTP phenotype as the as Fyn single mutant animals. These data suggest that:

- Yes is not important in LTP
- Src deficiency is compensated by Fyn
- Fyn deficiency cannot be fully compensated by Src

The cAMP pathway

Pharmacological studies on hippocampal slices have indicated that long-term LTP (L-LTP), a sustained potentiation of greater than 2 hours, requires:

- cAMP – an increase in cAMP levels is seen in the CA1 region after application of NMDA and, while cAMP antagonists block L-LTP, cAMP agonists facilitate L-LTP.
- New transcription – the application of transcription inhibitors after a tetanic pulse, or after the application of cAMP, blocks the induction of L-LTP. A critical phase of transcription is needed, 1 hour after the stimulus, that is, during the E-LTP phase.
- The synthesis of new proteins. Translation starts within the first hour of the onset of LTP expression, while E-LTP is still in progress.

A postulated pathway for the activation of L-LTP through the cAMP pathway is shown in Figure 8.3. cAMP, generated by adenylate cyclases, activates protein kinase A (PKA), which phosphorylates the cAMP response element binding protein (CREB) transcription factor, thus promoting association with the CREB-binding protein (CBP), the assembly of the transcriptional machinery, and the activation of cAMP-responsive genes. A recent transgenic experiment has shown that stimuli that generate L-LTP induce gene expression through the cAMP responsive element (CRE), a *cis*-acting element that mediates cAMP-responsive gene regulation following CREB binding. Mice were generated that bear a CRE-regulated reporter construct and *de-novo* gene expression induced by various forms of synaptic change was monitored by assaying the expression of the reporter in the hippocampus (Impey *et al.*, 1996). CRE-mediated gene expression was shown to increase after L-LTP, and this was blocked by inhibitors of PKA.

A knockout mouse has been generated in which the α and δ forms of CREB are deleted (Bourtchuladze *et al.*, 1994). These mice have demonstrated that CREB is crucial for the induction of L-LTP and for the conversion of short-term to long-term memory. LTP is initially normal, but decays to baseline 90 minutes after induction. Similarly, spatial learning is initially intact, but is lost after 30–60 minutes.

8.2.7 Molecules and memory – retrograde messengers

So far, we have considered changes in the postsynaptic neurone in response to a glutamate signal from the presynaptic axon. However, the maintenance of LTP requires a change in presynaptic function. How is the the postsynaptic induction event communicated to the presynaptic terminal? What is the nature of the retrograde signal? A number of candidates have been identified, including arachidonic acid, carbon monoxide and nitric oxide.

Nitric oxide

Nitric oxide (NO) is a noxious, highly reactive and diffusible free-radical gas with a short half-life. NO was first identified as a mediator of macrophage actions – macrophages produce nitrates that derive from NO, the precursor of which is arginine. Arginine derivatives, or arginine starvation, can be used to block the formation of NO, and these treatments block the tumoricidal and bactericidal actions of macrophages. Subsequently, NO has been shown to mediate blood vessel relaxation and immune responses and NO has been implicated as an important neurotransmitter in both the central and peripheral nervous systems

NO is derived directly from arginine by the enzyme NO synthase (NOS), which exists in a number of forms:

- inducible or macrophage NOS (iNOS)
- neuronal NOS (nNOS)
- endothelial NOS (eNOS)

Interestingly, nNOS and eNOS are dependent upon Ca^{2+} and calmodulin, and are phosphorylated by αCamKII and PKA (Figure 8.3).

A number of potential presynaptic NO targets have been identified (Figure 8.3), including soluble guanylyl cyclase, which is stimulated to generate cGMP, a second messenger which goes on to activate a phosphorylation cascade through a cGMP-dependent protein kinase (PKG). There is growing evidence that NO is synthesised postsynaptically and acts retrogradely on presynaptic axons to alter synaptic strength, through activation of axonal guanylyl cyclase. Again, pharmacological experiments have led the way, suggesting appropriate transgenic experiments. Thus, the use of NO scavengers, NOS inhibitors, and a membrane-impermeant NO donor that releases NO only upon photolysis with ultraviolet light, suggests that

NO is produced in the postsynaptic neurone, travels through the extracellular space, and acts directly in the presynaptic neurone to induce LTP. Similar experiments have suggested that guanylyl cyclase and PKG are the NO targets in the presynaptic neurone.

Suggestive though they are, these pharmacological manipulations of NO pathways have been criticised due to the non-specific effects of the drugs used. It was thus hoped that a transgenic approach would be less prone to artefact. However, the knockout of nNOS was initially something of a disappointment as far as brain function is concerned. The main phenotype observed was the development of a grossly distended stomach, with hypertrophy of the pyloric sphincter and circular muscle layer, but the CNS was normal and LTP and memory function were normal. However, it was noticed that the brains of the knockout mice continued to have a residual level of NOS activity, suggesting that other NOS genes were able to compensate for the absence of nNOS. Indeed, it was then shown that eNOS is expressed in CA1 hippocampal neurones. The next step was therefore to knock out the eNOS gene. In eNOS-deficient mice, LTP was again shown to be normal. However, in doubly deficient mice (Son *et al.,* 1996), null for both the nNOS and eNOS genes, LTP in the stratum radiatum (Figure 8.1) was reduced, but was normal in the stratum oriens (Figure 8.1). These data suggest that NOS is involved in LTP in the striatum radiatum, but not in the stratum oriens, and that nNOS and eNOS can both perform this function, and can compensate for each other. These results have been confirmed using an adenoviral approach that enables hippocampal slices to be efficiently transfected with eNOS mutants (Box 8.4).

BOX 8.4: NO SIGNALLING IN THE HIPPOCAMPUS – A VIRAL APPROACH

A truncated mutant of eNOS that lacks catalytic activity was shown to act as a dominant negative inhibitor of endogenous eNOS activity. This molecule, when introduced into hippocampal slices by adenoviral transfection, blocked LTP at synapses of the stratum radiatum, but not stratum oriens, suggesting a diversity of LTP mechanisms, even within the same neurone. The eNOS protein is myristoylated, and through this acyl group it associates with membranes. Myristoylation involves the co-translational addition of the fatty acid myristate to the N-terminal glycine of eNOS by *n*-myristoyltransferase (nNMT). nNOS is not myristoylated. In the presence of an nNMT inhibitor, eNOS is no longer membrane-bound, and LTP is attenuated. However, LTP can be rescued by the introduction of another modified eNOS. This time, eNOS was fused to the extracellular and transmembrane region of CD8, a cell surface marker. When eNOS is directed to the membrane by this myristoylation-independent mechanism, LTP is rescued in the presence of the nNMT inhibitor. These data support a role for eNOS in LTP in the stratum radiatum, suggest that membrane association is crucial for eNOS function in LTP, and indicate that it is eNOS, not nNOS, that primarily functions in LTP (Kantor *et al.*, 1996).

Figure 8.3 suggests molecular pathways that may be involved in the conversion of incoming information into LTP in the CA1 region of the hippocampus. It is also possible that these processes are involved in memory formation. Pre- and postsynaptic changes contribute to expression of LTP. How are these changes converted into LTP? Kinases enhance postsynaptic sensitivity to glutamate, and also enhance the presynaptic release of glutamate. However, there can be no presynaptic enhancement without an increase in postsynaptic Ca^{2+}, which thus demands some kind of retrograde messenger, perhaps NO. How do kinases alter the secretion of glutamate, and the functioning of NMDARs, in order to facilitate LTP? And what about phosphatases – do they have a role in LTP, learning and memory? (Yes, they do – see Mansuy *et al.*, 1998; Winder *et al.*, 1998.) Clearly, much remains to be learnt about these pathways, how they interact, and how they ultimately enhance synaptic strength.

There is growing evidence that LTP has characteristics that recommend it as being a mediator of learning and memory processes; however, this popular, almost axiomatic, concept is not without its critics. The difficulties inherent in interpreting gene manipulation studies on LTP, learning and memory have, quite rightly, been highlighted. Are the deficits observed a consequence of defects in synaptic plasticity, or are they due to spatially or developmentally distant effects? To a large extent, these criticisms have been addressed by the development of systems that enable inducible and cell-specific gene lesions.

In addition, deficiencies in particular experiments have been exposed. For example, the αCaMKII knockout mice are slow to learn the location of the hidden platform in a Morris watermaze, suggesting a role for this kinase in spatial learning. However, the deficient animals were slow to swim to the platform on the first training trial, before any learning could have occurred. The poor performance was attributed to 'fatigue' and 'jumpiness' in the mutant mice. The *fyn* knockouts displayed a similar performance deficit on the first trial, again before the point at which learning could have occurred. Furthermore, the mutants learned at a similar rate to the wild-type controls, and by the sixth trial, were performing at identical levels. In fact, because some mutant mice reached a preimposed 60 second cut-off on the first trial, the rate of learning for the group may have been underestimated, and may have exceeded that of the wild-type controls! Shors and Matzel **(http://www.cogsci.soton.ac.uk/bbs/Archive/bbs.shors.html)** have critically evaluated the evidence of a link between LTP and memory formation, and found it wanting. Rather, they propose that LTP may serve as a neural equivalent to an arousal or attention device in the brain. As such, LTP may facilitate the induction of memories at distant synapses.

8.2.9 Questions for future learning and memory research

1. Can we finally resolve the question of the relationship betwen LTP and memory formation?
2. What is the molecular basis for the physical storage of memories?
3. How are memories recalled at will?

8.3 Circadian rhythms

Unlike learning and memory, one aspect of behaviour that we may soon understand at a molecular level is **circadian rhythms.** Daily behavioural rhythms in activity and sleep do not passively follow the prevailing day/night cycle but are, in fact, controlled by an endogenous 'clock' mechanism (Box 8.5)

8.3.1 The brain's 'clock'

Circadian rhythms are timed by a brain 'clock' that also controls physiological rhythms. Precise lesioning techniques and tissue transplantation experiments have localised the brain's 24-hour clock to the suprachiasmatic nucleus in the hypothalamus. Similar techniques have yet to be applied to humans but an analogous structure in the human brain is thought to perform the same function. An expensive, although simple experiment to check your own endogenous clock is to take a trans-meridian (e.g. trans-Atlantic) flight and then go out for an evening meal with local friends. Pleasant as the experience may be, you will soon be aware of the activity of a timing mechanism that is not on your wrist.

As a discrete component of behaviour that is known to be generated within a defined group of neurones, the basis of neuronal rhythmicity has been perceived to be within our reach at a mechanistic and molecular level. Research into rhythms is not simply driven by the need to understand the phenomenon of jet-lag, but is of fundamental importance because daily rhythms are observed in most, if not all, species. (See Box 8.5. – analogous rhythmic behaviour/activity in simple organisms places rhythmicity in a special category of behavioural studies that can be related mechanistically to simple model organisms.) The study of rhythms is also medically important because experiments with rodents have shown that damage to the brain's clock is associated with altered behavioural rhythms, and perturbed rhythms are an inherited characteristic of particular genetic mutants. Clearly, a similar mutation could underlie specific human pathologies and, indeed, there is evidence to suggest that certain psychiatric conditions including some forms of depression, may be linked to disorders of endogenous timing. Taken together with the notion that perturbed sleep cycles could be rapidly re-set with a 'clock pill' (as distinct from a sleeping pill) it is not surprising that there is considerable medical and pharmaceutical interest in dissecting the workings of the brain's clock.

8.3.2 Taking the clock apart

The search for the molecular basis of the brain's clock has been at the forefront of recent experimental work in the molecular genetics of mammalian behaviour, and therefore provides a valuable reference point for similar investigations. This work also demonstrates the cohesive use of a multitude of new technologies discussed in earlier chapters: generation of mouse mutants, positional cloning, manipulation of large DNA clones, automated sequencing and bioinformatics, *in situ* hybridisation, and transgenic animal production. At the same time, the value of traditional approaches such as comparative species analysis is well illustrated by recent work in this field.

As in any dynamic field of research in which different groups of researchers are, in effect, competing in a race to obtain a particular result, the molecular basis of the clock has been approached in various ways. While the essential target of these studies – the identification of **clock genes** (necessary for clock function) – has been similar, the experimental approaches differ enormously in their difficulty and general utility. The most arduous approach – **forward genetic** (see section 3.4.2) which also probably has the greatest utility for behavioural genetics in general, has resulted in the successful cloning of what appears to be a clock gene, termed *Clock*. The forward genetic strategy (King *et al.*, 1997), was chosen because the only known spontaneous rhythm mutant is a hamster (*tau* – see Further reading and Liu *et al.*, 1997), a species that is not yet a viable model for positional gene cloning purposes. With respect to the availability of mouse mutants, it should be noted that circadian rhythmicity is limited in this respect because many other kinds of behavioural mouse mutants are long-established, and many can be obtained from the Jackson Laboratory (see section 3.5). New behavioural mutants are also being generated through programmes of random mutagenesis (see section 3.4.2).

The new mouse mutant *Clock* was identified through selective behavioural screening of offspring derived from a single male which had been subjected to **chemical mutagenesis** (using **ENU**; see Table 3.1) – in this study a total of only 304 first-generation offspring were screened for the desired phenotype. Not every forward genetic investigator looking for a specific dominant mutation could expect to be so fortunate because a saturation screen would involve ten times this number of offspring (see section 3.4. and Takahashi *et al.*, 1994). The rhythm mutant was also ideal for further study because *Clock* is a single locus mutation that affects both the duration, and the endogenous nature of the daily activity rhythm. Furthermore, the animals appeared normal with respect to other behaviours, indicating that the mutation was within a discrete component of the endogenous clock mechanism, and was not associated with a general inhibition of locomotor activity, for example.

The long road to identifying the genetic basis of the *Clock* mutation by positional cloning (Box. 3.1) was begun by localising *Clock* to a single chromosome. **SSLPs** were then used to map the gene more precisely to a chromosomal region, before both **SSLPs** and **STSs** from this region were used to isolate genomic clones – the essential resource that makes gene identification and sequencing possible. Both **YAC** and **BAC** libraries were used for clone isolation, and it is instructive to refer to the study of King *et al.* (1997) for further details of this process. BAC clones were used in a variety of extensive screening procedures that were designed to identify genes encoding *Clock*. Given that *Clock* had been mapped precisely to a defined chromosomal region, the following question should be asked at this stage – what potential short-cut was not taken by these researchers, and why? (see section 3.5). This is an important question because it is apparent from the published study that something like 2 million DNA bases were sequenced during the final stages of identifying the *Clock* gene! Nevertheless the exhaustive **'shotgun'** sequencing effort was eventually successful because the sequence of one clone was found to be homologous to a region of the *Drosophila* circadian clock gene *per* (Box. 8.5), and subsequent investigations provided further evidence that the mouse *Clock* gene had indeed been found. Had the conventional **homology screening** failed to provide a candidate clone, however, an additional approach used by these investigators confirmed that a particular BAC clone did indeed contain the *Clock* gene.

The additional approach used by the *Clock* gene team (Antoch *et al.*, 1997) has illustrated the principle of cloning by **transgenic rescue,** in which candidate DNA sequence isolated by positional cloning is used as a transgene to 'rescue', or complement, a mutant behavioural phenotype. Because the isolated clone is obtained from a wild-type library it can be assumed that the transgene will not contain the same functional mutation as the mutant animal. It should be noted that the two approaches used in the *Clock* study were essentially complementary because the large DNA clones such as BACs that are used as 'rescue' transgenes will, of course, eventually require extensive sequence analysis for the identification of functional transcription units. In a scenario where multiple candidate clones are available however, it may be wise to select clones by transgenic rescue before embarking on a large-scale sequencing effort. In the *Clock* study, one particular BAC clone was found to rescue the loss-of-rhythm phenotype of the mutant and, conveniently, subsequent analysis showed that only a single transcription unit was present in the clone. The presence of multiple transcription units (with or without knowledge of relevant gene homology) would demand individual rescue experiments. Other potential problems of inappropriate transgenic phenotype in rescue experiments can also be envisaged.

8.3.3 Putting the clock back together

The homology of the *Clock* gene (which exhibits a single point mutation in the Clock mutant; King *et al.*, 1997) to the *per* gene of *Drosophila* (Box 8.5) indicates that *Clock* might also function in a feedback loop oscillator mechanism. Has the identification of *Clock* therefore brought us close to an understanding of rhythmic behaviour in mammals? Hot on the heels of the *Clock* study, a different approach to the molecular basis of mammalian rhythmicity provided another mouse clock gene termed m*per* (Tei *et al.*, 1997) which exhibits even greater homology to Drosphila *per*. This gene was obtained through a direct attempt to clone mammalian *per* homologues using a novel RT-PCR (see section 1.2.2) technique called **IMS** (intra-module scanning)-**PCR** in which a variety of degenerate oligonucleotide PCR primers were designed acording to the sequence of a distinctive functional domain in *per*. Simultaneously, the human homologue of m*Per* was cloned by another research group who were simply surveying chromosome 17 specific transcripts; this illustrates the current power of DNA database searches (see Box 2.3) which can quickly provide researchers with insights into the potential function and biological importance of unknown sequences.

Thus, no sooner had *Clock* been isolated than another potential clock gene was identified, and, in fact, m*Per* mRNA oscillates in a robust circadian rhythm in the suprachiasmatic nucleus, unlike *Clock* which does not exhibit a distinct rhythm of expression. However, it is important to remember that *Clock* was isolated through the **functional screen** that comprises the forward genetic strategy, and confirmed as being necessary for circadian behaviour by transgenic rescue. Similar functional evidence is therefore required to show that m*Per* is a true clock component; currently it is not known whether m*Per* may be an example of a **clock-controlled gene** - a gene which oscillates within clock neurones but is not a required component of the clock mechanism. At the same time, doubts have also been raised concerning the possible role of *Clock* within the circadian clock mechanism. For example, Reppert and Weaver (1997) have pointed out that mice deficient in a particular neural cell adhesion factor (PSA-NCAM) exhibit a similar phenotype to the clock mutants - a finding which, together with other observations, has raised the possibility that the *Clock* gene product may act as a coupling molecule that strengthens the interactions between individual cellular clocks (see Box 8.5) and is not, in fact, an intrinsic clock component. These discussions illustrate the problems of interpretation that confront neuroscientists who are searching for molecular mechanisms of behaviour. At the present time, evidence in favour of an intrinsic role for *Clock* is gathering (see Gekakis *et al.*, 1998), but it seems likely that additional mammalian clock genes will be identified as the circadian clock mechanism is gradually asembled.

8.3.4 A different clock mechanism – antisense regulation?

The search for the molecular basis of circadian behaviour has taken another turn with some new findings derived from another model system – the giant silkmoth *Antheraea pernyi*. As described above, comparative studies of *Drosophila* and mouse have provided a convergence of interest on clock genes of the *per* family, and the *per* feedback loop model (Box 8.5) has been seen as a general mechanistic framework. However, recent publications on circadian rhythms in the silkmoth (Sauman and Reppert, 1996) have highlighted the value of comparative studies by showing that while the

BOX 8.5: MODELLING THE MOLECULAR MECHANISMS OF BEHAVIOUR – A TRANSCRIPTIONAL FEEDBACK MODEL OF CIRCADIAN CLOCK FUNCTION

All classes of organism from bacteria, to fungi, to moths, mice and man exhibit daily rhythms in activity that reflect the 24-hour day/night cycle of the solar environment. Analysis of rhythmic phenotypes in these diverse organisms has revealed that the rhythms are endogenous – they are maintained in experimental environments that are free of timing cues like sunrise, and are lost, in rats for example, following discrete lesions of the brain. These endogenous daily rhythms are termed **circadian.** The conservation of circadian rhythms throughout evolution indicates that the capacity to generate and maintain (roughly) 24-hour rhythms – the phenomenon of the so-called **biological clock** – is a property of single cells. This appears to be true even for a complex nervous structure such as the mammalian brain because the brain's clock – the **suprachiasmatic nucleus** – appears to be composed of a multitude of synchronised, cell-autonomous, neuronal clocks (Welsh *et al.*, 1995; Liu *et al.*, 1997).

Exactly how individual cells can function as tiny clocks is an intriguing question for biologists, and it is therefore not surprising that many different experimental models have been investigated. In recent years, a widely accepted model of a circadian clock mechanism has been pieced together from genetic studies on both the fruit fly *Drosophila*, and a less widely used model, the fungus *Neurospora*. Using these relatively simple organisms it has been possible to identify genes that are crucial for circadian clock function. Most recently, mammalian homologues of 'clock' genes have also been identified (section 8.3.3). The 'chronobiologists' responsible for these important breakthroughs have formulated a molecular mechanism of cellular clocks that can be schematised as shown in the Figure.

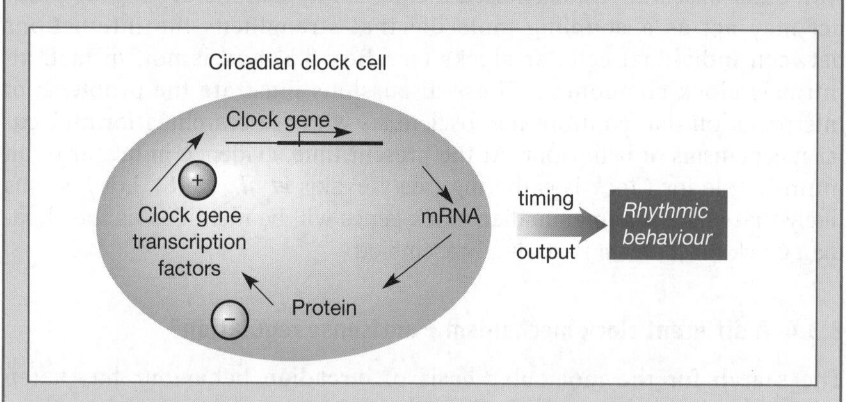

Hypothetical model – an autoregulatory transcriptional feedback loop clock

In this model, the **clock gene** which determines circadian behaviour is an intrinsic molecular component of the mechanism, as distinct, for example, from a gene encoding a neuropeptide which may simply modulate an aspect of a particular behaviour. The rhythm of clock gene transcription is hypothesised to be self-

BOX 8.5: Contd

sustaining through a negative feedback action of the gene product on transcription factors which positively regulate clock gene transcription (see Dunlap, 1998). This molecular activity is thought to be linked to a cellular output mechanism which, in turn, is linked to the control of circadian behaviours. Consider the properties of the clock protein(s) that are required for this model to function. How would you test the validity of the model? What alternative models are consistent with the available data (see Further reading)? Also refer to the Biological Clocks home page (**http://www.cbt.virginia.edu/bio419/clocks1.htm**) for more information on circadian clocks, and links to other timing-related web sites.

PER protein of this species does indeed appear to control certain circadian behaviours, the molecular mechanism may be quite distinct.

Firstly, it was shown that the circadian clock neurones in the silkmoth brain do not exhibit a nuclear rhythm of PER protein – a transcriptional feedback model would therefore not appear to be tenable in this organism. Why not? Secondly, *in situ* hybridisation analysis of *per* transcript expression in the silkmoth brain using a sense RNA probe (initially as a negative control!) revealed a rhythm of **antisense** transcript expression – a rhythm which was similar in amplitude to, but temporally opposite to that of the *per* mRNA (sense) circadian rhythm. The authors of this study have suggested that *per* antisense transcripts regulate sense transcript levels, and hence the rhythmic expression of PER protein – a suggestion which is consistent with recent insights into the biological role of eukaryotic antisense RNA. What could be an alternative interpretation of 'antisense' RNA expression, and how could this alternative be investigated? These findings have shown that the same gene *(per)* may control circadian rhythms in different species, but possibly via diverse mechanisms – **mechanistic pleiotropy** (Sauman and Reppert, 1996). Antisense regulation of mammalian circadian clocks (and other behaviours) should also now be investigated.

8.4 Summary

- **Learning** and **memory** can be studied in rodents using **memory tests** such as the **Morris watermaze**.
- The **hippocampus** appears to be involved in the conversion of **short-term** memories into **long-term** memories.
- **Long-term potentiation (LTP)**, the activity-dependent potentiation of synaptic efficiency, has characteristics that recommend it as a molecular substrate for learning.
- Candidate learning and memory genes have been modified in **transgenic mice**.

- Problems with **germline transgenesis** have been overcome using **inducible** and **cell-specific** transgenic systems.
- Hippocampal LTP is mediated by **NMDA receptors**.
- Hippocampal LTP involves the Src family of tyrosine kinases.
- Hippocampal LTP involves α**CamKII**.
- Hippocampal LTP involves the **retrograde** messenger **nitric oxide**.
- Hippocampal long-term LTP involves the **cAMP** pathway.
- Genetically mediated **LTP** deficits correlate with deficits in learning and memory.
- Daily rhythms in behaviour are termed **circadian**, and are controlled by a cellular clock mechanism located in the suprachiasmatic nucleus of the brain.
- The molecular basis of the clock has been investigated using a forward genetic strategy in mice that involves testing for behavioural mutants derived by chemical mutagenesis. A mutant **clock gene** has been identified by positional cloning.
- Replacement of the wild-type **clock gene** using transgenesis corrected the mutant behavioural phenotype – an approach termed **transgenic rescue**.
- The molecular mechanism of circadian timing is hypothesised to involve either: (i) a feedback mechanism in which a clock gene product autoregulates transcription in a temporally-limited manner; or (ii) the generation of antisense transcripts which temporally limit clock gene expression.

Further reading

GRANT, S.G.N. and SILVA, A.J. (1994) Targeting learning *Trends in Neuroscience*, **17**, 71–5.

A good summary of the 'first generation' of knockout mice that were used to ask questions about LTP, learning and memory.

MALENKA, R.C. and NICOLL, R.A. (1997) Never fear, LTP is hear. *Nature*, **390**, 552–553.

A Nature *'News and Views' article describing experiments that clearly strongly suggest that LTP is involved in memory processes. Animals were trained to associate a tone with a foot shock, eliciting a fear-conditioning response. This causes an increase in the strength of synapses in a region of the brain that is associated with learning and memory. After training, this response occurred when the tone was given but the foot shock was not.*

MILNER, B., SQUIRE, L.R. and KANDEL, E.R. (1998) Cognitive neuroscience and the study of memory. *Neurone*, **20**, 445–468.

As good as it gets. This definitive review puts recent molecular studies in a broader historical context and asks, 'can ...two independent and disparate strands – molecular neurobiology and cognitive neuroscience – be united'?

TAKAHASHI, J. (1995) Molecular neurobiology and genetics of circadian rhythms in mammals. *Annual Review of Neuroscience*, **18**, 531–553.

A comprehensive review at the interface of traditional and molecular approaches to circadian clock function.

HALL, J. (1995) Tripping along the trail to the molecular mechanisms of biological clocks. *Trends in Neuroscience*, **18**, 230–240.

The title says it all – a useful illustration of the reality of scientific progress.

CHAPTER NINE

Neurodegenerative diseases

Key topics

- Neuronal death
    - Apoptosis
    - Necrosis
    - Selective vulnerability of neurones
- Alzheimer's disease
    - Amyloid plaques and β-amyloid
    - Neurofibrillary tangles and tau
    - Alzheimer's disease genes
    - Transgenic models
    - APP frameshift mutants
- Prion diseases
    - Mad cow disease (BSE)
- Huntington's disease
    - Trinucleotide and glutamine repeat diseases
    - Loss-of-function and gain-of-function mechanisms
    - Transgenic models
    - Neuronal intranuclear inclusions

9.1 Introduction

Functional deterioration and eventual loss of neurones (**neurodegeneration**) characterises a number of adult-onset neurological disorders such as:

- Alzheimer's disease (AD, 7–10% of everyone over 65 years).
- Parkinson's disease (PD, 1% incidence at 65 years).

- Huntington's disease (HD, 0.01% incidence in Western European populations).
- Spinal and cerebellar ataxia type-1 (SCA-1, rare).
- Machado–Joseph disease (MJD, SCA-3, rare).
- Spinal and bulbar muscular atrophy (SBMA, rare).
- Dentatorubral-pallidoluyian atrophy (DRPLA, inherited form is prevalent in Japanese).
- New variant Creutzfelt–Jacob disease (vCJD, predicted incidence from just 75 cases to 80000).

Many of these diseases have now been distinguished into multiple variants that include both **familial** (inherited) and **sporadic** (non-inherited) forms. Certain diseases including vCJD are transmissible and appear to involve infectious agents termed **prions** (Box 9.1). Some diseases show convergence between sporadic and familial forms which provides an indication of a common underlying mechanism.

Neurodegenerative diseases may be likened to other contemporary diseases such as diabetes and cancer in the sense that they are now generally perceived as major health-care problems – even before the recent 'BSE crisis' (Prusiner, 1997). Much current molecular neuroscience research is directed towards an understanding of the aetiology of these disorders, which will be discussed here with reference to selected examples. A wide-ranging debate involving eminent neurodegenerative disease researchers has recently appeared in the BioMedNet on-line magazine, HMS *Beagle* (see **http://biomednet.com**).

BOX 9.1: YOU ARE WHAT YOU EAT ... MAD COW DISEASE AND PRION PROTEINS

Several rare neurodegenerative diseases including kuru, Creutzfeldt–Jakob disease (CJD), Gerstmann–Straussler–Scheinker disease (GSS) and fatal familial insomnia (FFI) are now recognised as **prion** ('protein infectious agent') diseases. These transmissible spongiform encephalopathies (**TSEs**) are characterised by a specific pathology – vacuolisation of neuronal cytoplasm which results in a typical sponge-like appearance of the brain – and various neurological symptoms that include ataxia and dementia. The nature of the agents which transmit these diseases has been, and still is, the subject of intense debate and research conflict (see Aguzzi and Weissman, 1997), but most scientists in this field are now agreed that prions are involved. The molecular mechanisms through which prions cause neurodegeneration are far from defined, although the currently accepted story is that a normal brain protein PrP^{C} (prion protein, normal cellular form) can be converted into a misfolded, pathological form, PrP^{Sc} (scrapie prion protein form; **scrapie** is a related disease of sheep) by interacting with exogenous PrP^{Sc}, hence the transmission/'infection'. Following this initial interaction and conversion, an exponential conversion cascade is hypothesised to occur:

BOX 9.1: Contd

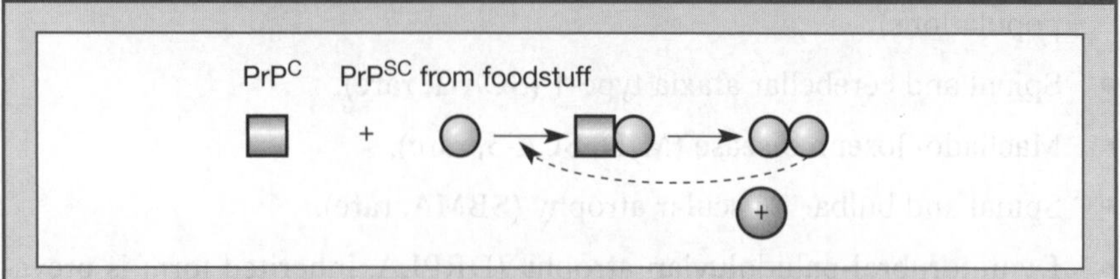

The key aspect of the transmission process is that it can be oral – the protease-resistant prions in foodstuff eventually reach the brain and initiate the neurodegenerative disease. In the past, this scenario has led to either isolated outbreaks of disease, such as the kuru epidemic in Papua New Guinea where the disease appears to have been propagated by ritual cannibalism, or isolated cases of CJD transmitted by injection of prion-infected pituitary hormones to endocrine patients (before the development of recombinant hormones). Most recently, however, a form of non-ritual cannibalism that involved the feeding of contaminated cattle offal to cattle has propagated an outbreak of an animal prion disease termed **bovine spongiform encephalopathy** (**BSE**; mad cow disease; see Prusiner, 1997). Alarmingly, it now appears that BSE can in turn be transmitted to humans, resulting in the appearance of a new variant of CJD (**vCJD**) that is characterised by early onset (below the age of 35) and novel neuropathological features. Transmission of prion diseases between species was denied by the UK authorities until experiments on mice (for example, see Bruce *et al.*, 1997) provided clear experimental evidence. The full extent of the vCJD epidemic has yet to be realised, and estimates vary widely from just 75 cases to 80 000 (see the 'mad cow disease' web site: **http://www.mad-cow.org/index.html** and the web site of the UK CJD surveillance unit: **http:www.cjd.ed.ac.uk/index.htm**).

9.2 Neuronal death

The loss of neurones in these diseases implies the involvement of a 'cell death' mechanism. Cell death can be distinguished into two major subtypes:

- Programmed (**apoptosis**)
- Non-programmed (**necrosis**)

Programmed cell death is generally perceived as 'naturally occurring' cell loss that is seen, for example, during nervous system development when as much as 85% of some neuronal populations die away during developmental organisation. Non-programmed or necrotic cell death has, until recently, been thought to represent the adult-onset loss of cells in disease states. However, it is now apparent that apoptosis, which is an active process involving gene induction, also occurs in pathological situations such as the neurodegenerative diseases.

The relationship between apoptosis and other mechanisms which have been linked to cell death – **oxidative stress** and **excitotoxicity** – is unclear, and has been an active area of research. However, with respect to the neurodegenerative diseases, there is currently limited experimental support for underlying oxidative and excitotoxic mechanisms of cell death.

Currently, attention is focused on the role of abnormal protein deposits that are a common feature of the neurodegenerative diseases (Lansbury, 1997). It is apparent that these diseases are essentially '**proteinopathies**' – whereby disease-related protein complexes interfere with cellular function, and although a causal link has not been established, cell death is an ensuing event. The generation of toxic protein complexes could potentially arise through a variety of mechanisms including:

1. Disease mutation generates a novel toxic protein product.
2. Over-expression of disease gene (transcriptional or post-transcriptional).
3. Altered metabolism of disease protein – a novel proteolytic event may generate a free (and thereby toxic) domain.
4. Novel cellular compartment – e.g. nuclear expression in disease state only – may result from (3).
5. Novel protein interaction – as a result of (3) and/or (4) the disease protein may interact with another protein, leading to toxicity through either altered conformation (protein folding), or interference with 'partner' protein function.

9.2.1 Selective vulnerability of neuronal subgroups

The different symptoms of the various neurodegenerative diseases reflect neuronal loss within specific cellular populations: for example, striatal neurones in HD and nigral neurones in PD. However, this is not due to cell-specific expression of particular disease genes, some of which are expressed ubiquitously (e.g. see HD gene below). A key question therefore concerns the factors which determine the selective vulnerability of subsets of neurones.

Regarding some of the mechanisms which have been proposed to underlie neurodegeneration (**excitotoxicity** and **oxidative stress**), it has been suggested that cell loss is largely restricted to neurones which have some special vulnerability to these factors. There may be some validity to this argument (reduced expression of the gluR2 glutamate receptor subunit gene may confer excitotoxic sensitivity; Pellegrini-Giampietro *et al.*, 1997), but in general these factors are not known to be selective for subsets of neurones.

With respect to the potential mechanisms linked to toxic protein deposits (see section 9.2), cell-specificity could arise in a number of ways:

- Cell-specific expression of mutant protein.
- Cell-specific over-expression of disease gene (altered synthesis or turnover).

- Cell-specific proteolytic modification of disease protein.
- Cell-specific, subcellular localisation.
- Cell-specific protein : protein interaction.

Studies are currently underway to investigate the utilisation of such mechanisms in neurodegenerative diseases, including Alzheimer's and Huntington's disease.

9.3 Alzheimer's disease

Alzheimer's disease (AD) is the most common form of dementia in the elderly, affecting some 15–20 million people world-wide. It is a devastating condition for both patient and family, being associated with a progressive loss of memory and other cognitive functions that mirrors massive neurodegeneration. (Images of brain scans showing a comparison of AD and normal brains can be viewed at: **http://teri.bio.uci.edu/dement.html**.) AD is commonly sporadic, but recent genetic studies have highlighted inherited forms of the disease, and AD susceptibility genes have been identified (see section 9.3.3).

In an increasingly ageing population (within the developed world) AD is set to become one of the major healthcare problems of the 21st century. Currently there is no cure, or even effective treatment. Research into the basic mechanisms of AD is therefore of high priority, and accordingly represents an increasing proportion of neuroscience research. A general web site for AD research that includes 'hot papers' and a list of AD researchers world-wide can be found at: (**http://dsmallpc2.path.unimelb.edu.au/ad.html**). An on-line AD journal is located at: **http://www.coa.uky.edu/ADReview/**.

9.3.1 Pathological features of AD

Post-mortem examination of AD brains reveals a characteristic pathology that includes the presence of both abundant extracellular **amyloid plaques/deposits** (similar 'senile plaques' are also observed in non-Alzheimer aged brains), and intracellular **neurofibrillary tangles/lesions**.

These structures, which are both composed of fibrous proteins, are believed to interfere with cellular and synaptic function, leading to degeneration (see section 9.2). A number of different neuronal systems exhibit neuronal dysfunction and death, but cholinergic neurones of the basal forebrain are primarily affected in the majority of investigated cases. The fibrous structures provide an obvious starting point for an analysis of the molecular basis of the disease and, in fact, identification of the constituent proteins has caused AD researchers to become divided into two groups:

1. The 'tauists', who study the **tau** protein which is a component of the intracellular fibrils.

2. The 'βaptists' who study the β **amyloid** peptide Aβ1–42 which is the major component of the extracellular fibrils.

The religious denominations (Beyreuther and Masters, 1996) imply a certain degree of controversy in this research field with regard to the primary cause of the disease.

Tau filaments

The intracellular fibrils are composed of two kinds of abnormal filaments:

1. Paired helical filaments (PHFs)
2. Single straight filaments (SSFs)

Multiple isoforms of the protein tau are found in both types of filament, and because tau is a normal (axonal, microtubule-associated) cellular protein it was assumed that neurofibrillary tangles (located in cell bodies and apical dendrites) may be composed of a mutant form of tau. This is not the case – but it has been shown that the PHF tau in AD brains is **hyperphosphorylated**, and a consequent reduction in binding to microtubules may cause cellular dysfunction through the loss of a vital microtubule-associated process such as rapid axonal transport. Recent studies by Michel Goedert and colleagues have now provided evidence that the assembly of tau into fibrils requires an additional factor (**glycosaminoglycans** or **GAGs**) which are indeed associated with neurofibrillary tangles (see Beyreuther and Masters, 1996).

β-amyloid plaques

Mature β-amyloid plaques exhibit an abundance of the peptide Aβ1–42(43) (commonly termed Aβ), which is a carboxy-terminal proteolytic product of the amyloid precursor protein (APP). The proteolytic mechanisms that determine Aβ production have been illuminated by the study of AD genetics (see section 9.3.2). A large proportion of secreted amyloid peptides are the soluble 40-amino acid form, whereas Aβ1–42 is deposited and becomes aggregated into the characteristic extracellular plaques which are resistant to proteolytic digestion. It has been suggested that Aβ is the principal causative factor in the formation of plaques through the seeding of protein polymerisation (Lansbury, 1997). Over-expression of APP, as in **Down's syndrome**, where there are multiple copies of the (chromosome 21) APP gene has also been shown to be sufficient for amyloid deposition and neuropathology which is similar to AD. The mechanism by which Aβ leads to neurodegeneration is undefined. Recently, a **yeast two-hybrid** protein interaction screen has identified a neuronally enriched endoplasmic reticulum (ER) associated binding protein (ERAB) which binds to Aβ, and is implicated in neurotoxic mechanisms (Yan *et al.*, 1997) although this finding requires confirmation.

9.3.2 AD genes

Genetic studies have now identified a variety of gene mutations associated with AD that permit the following sub-classification of the disease:

Type of AD	*Mutant/susceptibility gene*
AD1; familial, autosomal dominant	*APP*
AD2; late-onset, familial, sporadic	*apolipoprotein E* (*apoE*)
AD3; familial, autosomal dominant early-onset	*presenilin 1* (*PS1*)
AD4; familial, autosomal dominant	*presenilin 2* (*PS2*)

The three autosomal-dominant familial mutations are relatively rare, accounting for <1% of AD, but the identification of these mutations has been valuable for AD researchers because it has provided insights into the mechanisms of β-amyloid plaque formation. Thus, mutations in the *APP* gene are located near recognition sites of the secretase enzymes that cleave APP, and are associated with enhanced production of Aβ. Similarly, the product of the *presenilin 1* gene is thought to regulate an APP-cleaving secretase, and cells of PS1 knockout mice are deficient in Aβ generation (De Strooper *et al.*, 1998) whereas transgenic mice expressing a mutant PS1 exhibit accelerated amyloid deposition (Borchelt *et al.*, 1997).

In contrast, the presence of the e4 allele of the *apoE* locus is thought to account for >50% of the susceptibility for AD. The role played by the mutant apoE product in the pathogenesis of AD is currently unclear, although there is a recent indication of a relationship with age of onset of the disease (see Price *et al.*, 1998). Future studies will be directed towards an ordering of the functional associations between these four (and additional!) gene products until the molecular cascade of AD pathology is established.

9.3.3 Transgenic animal models of AD

Apart from the models mentioned in section 9.3.2, many investigators have attempted to derive a general animal model for AD that would enable definitive studies of the pathological progression of the disease, and also permit testing of potential therapies. A major concern has been to investigate early events in the pathological process, before major neurodegeneration has occured. Such analysis in animal models may identify candidate therapeutics which could prevent the progression to major neuronal loss. Alternatively, treatments could be developed for pathological features of AD other than cell loss which may truly underlie the dementia (see Guest Box overleaf).

9.3.4 Frameshift mutants of APP

Molecular neuroscience, like all research fields, often progresses by drawing from discoveries in apparently unrelated areas of study. An excellent example of the importance of basic neuroscience research to the study of disease

GUEST BOX BY P. CHAPMAN

Alzheimer's disease in mice?

Alzheimer's disease presents a particularly thorny set of problems for those who hope to discover a cure. Its effect on behaviour is well known; the most obvious and devastating is the progressive and irreversible decline in the ability to think and reason (dementia). Inevitably, Alzheimer's disease is fatal, but it is only on autopsy that the amyloid plaques, neurofibrillary tangles and neuronal cell death that are characteristic of the disease can be confidently diagnosed. Therefore, our ability to understand the earlier stages of the disease by studying humans is compromised.

The obvious solution to this problem is to develop an animal model. By reproducing the pathology of Alzheimer's disease in a mouse or rat, we can hope to discover which features are responsible for the onset and progress of the disease, which are the proximal causes of dementia, and which are epiphenomenal (i.e. appearing with the critical pathologies, but not themselves pathological). I have been fortunate enough to be part of a team studying a transgenic mouse model developed by Karen Hsiao. These mice carry a mutated version of the human gene for amyloid precursor protein (APP), which has been implicated in Alzheimer's disease both biochemically (as the precursor for β-amyloid) and by genetic linkage analysis. The mice carrying the transgenes are indistinguishable from their non-transgenic littermates when they are born, and as they go through adolescence into adulthood. As they age, however, the transgenic mice begin to show characteristics of Alzheimer's; Steven Younkin (Mayo Clinic) showed that brain concentrations of β-amyloid are increased, while Greg Cole (UCLA) and Brad Hyman (Massachussetts General Hospital) demonstrated deposition of amyloid plaques in the forebrain. Karen Hsiao and I discovered that the mice's ability to learn and remember decreases at the same time the other pathologies are developing. By the time they are 18 months old they have the same number and distribution of amyloid plaques as humans in advanced stages of Alzheimer's, their brain β-amyloid concentrations equal those of humans with Alzheimer's, and they seem incapable of learning certain important behavioural tasks. To my mind, the big questions are: (i) how can we be certain these mice have something like Alzheimer's disease?; and (ii) how can we use the mice to further our understanding of Alzheimer's aetiology? The answers to these questions will not come quickly, but we are beginning to address them both. The neuropathological and behavioural data suggest many features of Alzheimer's, including age-dependent onset. Perhaps more exciting, through this model we have found that the behavioural and chemical abnormalities are associated with deficits in synaptic plasticity, long thought to be responsible for learning and memory. Moreover, all these symptoms have come in the absence of either neurofibrillary tangles or extensive neuronal cell loss, suggesting that some of the features we think are central to the disease may not cause dementia, while others we would not intuitively associate with it will provide the key targets for our attempts to cure Alzheimer's.

GUEST BOX Contd

HSAIO, K.K., CHAPMAN P.F., NILSEN S., ECKMAN C., HARIGAYA, Y., YOUNKIN, S., YANG, F. and COLE, G. (1996) Correlative memory deficits, Aβ elevation and amyloid plaques in transgenic mice. *Science*, **274**, 99–102.

Paul Chapman began studying the neurobiology of learning and memory in Richard Thompson's laboratory at Stanford University, where he received his PhD. After post-doctoral work at Yale University, he set up his own small but deeply disturbed laboratory in the Psychology Department at the University of Minnesota. While studying the basic mechanisms of synaptic plasticity in the mammalian forebrain, he began attempting to link synaptic plasticity, learning and Alzheimer's disease, in collaboration with Karen Hsiao at the University of Minnesota. He is now continuing this work in the Cardiff School of Biosciences.

mechanisms has recently been demonstrated by Fred van Leeuwen and colleagues in Holland who have demonstrated mutant forms of APP *transcripts* in AD patients (van Leeuwen *et al.*, 1998). These transcripts, which lack two bases in a coding region of the APP mRNA and are therefore **frameshift mutants**, derive from normal genes apparently through either a transcriptional or post-transcriptional *error* mechanism. Accumulation of the resultant mutant proteins (termed +1 proteins) in the

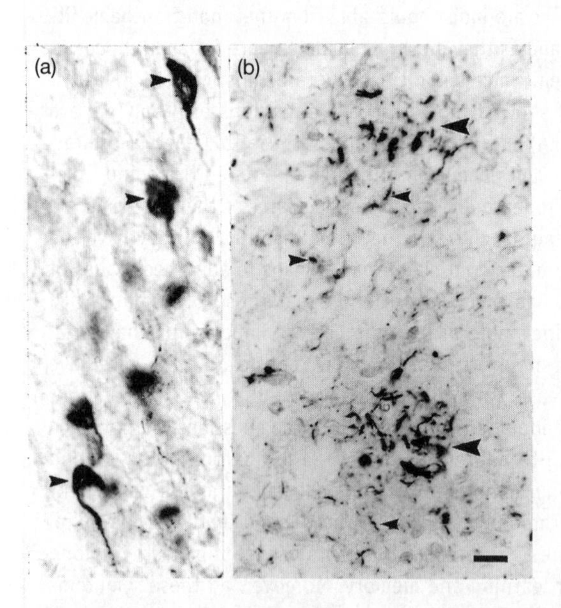

Figure 9.1 +1 proteins (**a**) ubiquitin-B^{+1} and (**b**) β-amyloid precursor protein^{+1} in sections of the cerebral cortex of two Alzheimer patients. Note that the neuropathological hallmarks of Alzheimer's disease are labelled: in (**a**), neurofibrillary tangles (small arrowheads) and in (**b**), the dystrophic neurites (large arrowheads) that form the neuritic plaques and the neuropil threads [small arrowheads in (**b**)] between the plaques. (**b**) Scale bar = 20 μm.

deposits of AD brains (see section 9.3.1) could contribute to AD pathology, and a co-localised error in the synthesis of ubiquitin (which functions in aberrant protein degradation) would exacerbate the problem (Figure 9.1). It is apparent that this mechanism could be an important factor in sporadic cases of AD. This important discovery was made because the Dutch research group were building upon their earlier finding of mutant transcripts derived from another gene in rats (Guest Box p. 35). Interestingly, the frameshift errors in rats were observed in neurones during ageing, and this type of mutation may therefore provide a link between ageing and neurodegenerative diseases.

9.4 Huntington's disease

Huntington's disease (HD) is an autosomal-dominant disorder characterised by abnormal spontaneous movements termed **chorea**, as well as emotional and cognitive impairment. Although it is generally an adult-onset disease, symptoms can appear during childhood. There is no effective therapy.

9.4.1 Trinucleotide (triplet) repeat expansion diseases

During the cloning and sequencing of many disease genes in the early 1990s, a distinct group of genetic diseases became apparent which are characterised by expansion of a trinucleotide sequence (e.g. CAGCAGCAG, etc.) in affected individuals – see: **http://www.biochem.emory.edu/warren/Trinucleotide.html**. These diseases are related with respect to a common underlying molecular pathology, but include diverse diseases such as myotonic dystrophy, fragile X syndrome and HD; a sub-classification into Type 1 (CAG repeats encoding polyglutamine; see following section) and Type 2 (non-CAG repeats outside the protein-coding region of the gene) has been made. An important characteristic of the trinucleotide repeat diseases is the phenomenon of **genetic anticipation** – earlier ages of disease onset in successive generations, and also increasing severity of symptoms. The molecular mechanism underlying this phenomenon appears to be **DNA instability** that can give rise to expansion of the nucleotide repeat in subsequent generations. Anticipation is also observed in other inherited diseases which have an unknown genetic basis, hinting that the relevant disease gene may contain a similar unstable mutation.

Glutamine repeat neurodegenerative diseases

These are a subgroup of the trinucleotide repeat diseases which include HD and a number of others identified to date (Figure 9.2). As the name implies, the trinucleotide repeat in these genes (CAG) is within the coding region of the gene and generates a **polyglutamine** domain which is expanded in HD. Not all trinucleotide repeats are coding – an expanded CGG repeat in the 5' untranslated region of the *FMR1* mRNA of fragile X syndrome acts to suppress translation (Feng *et al.*, 1995).

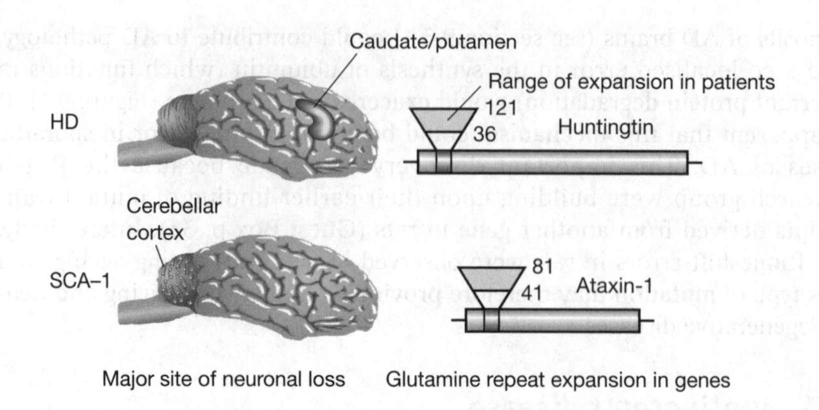

Figure 9.2 Examples of glutamine repeat diseases. Huntington's disease (HD) and spinal cerebellar ataxia type 1 (SCA-1). An expanded CAG repeat within a disease-specific gene is translated into a polyglutamine repeat region within the protein. Although the disease genes are widely expressed, neurodegeneration is localised to specific brain regions in a pattern that is characteristic of the particular disease.

9.4.2 Molecular pathology of Huntington's disease

Despite the fact that it is a relatively rare disease, HD and the mechanisms that underlie the associated neurodegeneration have been studied intensely. This focus is justified because it is recognised that the glutamine repeat neurodegenerative diseases may well share a similar mechanism of pathogenesis, and thus the insights gained from the study of HD can be applied generally.

Although the HD gene was cloned some years ago through the efforts of an international team (HD Collaborative Research Group, 1993), this important breakthrough did not provide any immediate clues about the mechanism of selective neurodegeneration. Thus, although the HD gene in affected individuals exhibits an expanded CAG repeat (36 to 121 repeats) as compared with 6–35 in normal individuals, the gene is expressed in a ubiquitous fashion that is not clearly altered by the CAG expansion (see MacDonald and Gusella, 1996). Also, the HD protein (**huntingtin**) sequence does not display significant homology to known proteins and therefore the normal function of huntingtin cannot be guessed (see Box 2.3 for details of protein sequence homology searches).

Modelling Huntington's disease in mice

Analysis of several aspects of HD and the HD gene including disease inheritance and the effects of null mutations in mice strongly suggest that the neurodegeneration is associated with a gain-of-function (see Box 9.2). This hypothesis has now received support from a study showing that '*knocking-in*' an expanded CAG repeat into the endogenous mouse (*Hdh*) gene does not impair huntingtin's normal (developmental) function (White *et al.*, 1997). But does this mouse line exhibit HD-type pathology? At the time of writing this is not clear because the mice are still under observation for an age-related phenotype (see White *et al.*, 1997).

BOX 9.2: ANALYSIS OF MOLECULAR PATHOLOGY: LOSS-OF-FUNCTION OR GAIN-OF-FUNCTION?

Where a disease mechanism is unknown, as in the example of Huntington's disease (HD), the available evidence can be used to ask whether the mechanism involves:

- a **loss-of-function**: the disease mutation impairs either the normal expression or function of the disease gene; or
- a **gain-of-function**: the disease mutation does not cause the pathology through interfering with the normal function of the disease gene – rather, a deviant property/function is generated by the mutation, resulting in the pathology.

Both clinical evidence and experimental animal data can be used to distinguish between these two alternatives. The following list summarises some relevant variables:

Loss-of-function	*Gain-of-function*	*Example of HD*
Recessive inheritance[1]	Dominant inheritance	Dominant inheritance
Deletion of one allele causes (partial) pathology	Deletion of one allele does not cause pathology	Deletion of one allele (Wolf–Hirshhorn syndrome) does not cause HD [2]
Homozygotes exhibit (full) pathology	Homozygote pathology is similar to heterozygote	Homozygote HD patients are similar to heterozygotes [2]
Genetic mutation can stop ORF	Genetic mutation does not stop ORF	HD protein is altered only by expanded polyglutamine
Protein function is reduced or lost	Protein function is modified	Expanded HD polyglutamine exhibits enhanced binding to another brain protein [3]
Gene knockout in mice causes disease pathology	Gene knockout in mice causes unrelated phenotype	Knockout of mouse HD gene is embryonic lethal[4]

On considering this analysis, it should become apparent that a distinction between either loss- or gain-of-function is overly simplistic, and open to different interpretations (see Martin, 1995). Nevertheless, this classification of disease mechanisms is widely used, and serves as an important aid to data interpretation. Try to apply the listed criteria to other neurodegenerative diseases using data available in the literature.

[1] Some loss-of-function diseases such as retinoblastoma may exhibit dominant inheritance – a second mutation in this growth suppressor protein can occur during somatic cell division. *See* MARTIN, J.B. (1995) CNS genetic disorders: loss of function, gain of function, or something else. *Current Opinion in Cell Biology*, **5**, 669–673.

BOX 9.2: Contd

[2] MacDonald, M.E. and Gusella, J.F. (1996) Huntington's disease: translating a CAG repeat into a pathogenic mechanism. *Current Opinion in Neurobiology*, **6**, 638–643.

[3] Li, X-J., Li, S-H., Sharp, A.H., Nucifora, F.C., Schilling, G., Lanahan, A, Worley, P., Snyder, S.H. and Ross, C.A. (1995) A huntingtin-associated protein enriched in brain with implications for pathology. *Nature*, **378**, 398–402.

[4] Duyao, M.P., Auerbach, A.B., Ryan, A., Persichetti, F., Barnes, G.T., McNeil, S.M., Ge, P., Vonsattel, J.P., Gusella, J.F., Joyner, A.L. and MacDonald, M.E. (1995) Inactivation of the mouse Huntington's disease gene homolog *Hdh*. *Science*, **269**, 407–410.

A broader question, which has now been addressed using transgenic technology, concerns the importance of **gene context** in the pathological mechanisms of the glutamine-repeat neurodegenerative diseases. To what extent are the disease genes simply carriers for the expanded CAG mutations? Is a pathology observed in a different gene context? The results of two different transgenic models are illustrated in Figure 9.3. It is apparent from these models that an expanded polyglutamine domain can cause a neurological phenotype irrespective of the gene in which the mutation is expressed. However, the *cell-specificity* of the neurodegeneration (see Figure 9.2) in the different expanded polyglutamine diseases would appear to depend upon the sequence context in which the polyglutamine is expressed. Future observation of the knock-in CAG mice (White *et al*., 1997) will be interesting in this respect. Given the available evidence, what are other possible explanations for cell-specific neurodegeneration?

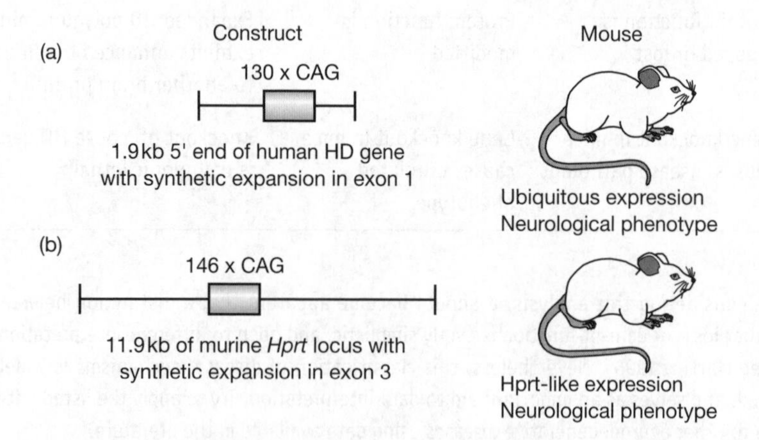

Figure 9.3 Transgenic mouse models of Huntington's disease (HD). Two of the approaches used to investigate the molecular basis of HD pathology have involved (**a**) random integration of a fragment of the human HD gene into the mouse genome and (**b**) targeted integration of a CAG repeat into a specific endogenous mouse gene (*Hprt*).

How do expanded polyglutamine-repeats cause neuronal dysfunction and degeneration?

Although a gene context-independent polyglutamine toxicity has now been identified in transgenic mice (see previous section), the biochemical mechanism underlying the mutant phenotype has not been defined. As discussed above (see section 9.2.1), cell-specific dysfunction could arise through a number of mechanisms. What is the evidence for these mechanisms in HD and expanded polyglutamine mouse models ?

As shown on p.153, the mutated huntingtin of HD is altered only by the presence of an expanded polyglutamine repeat, and this protein is expressed in all tissues. Furthermore, mutated huntingtin is expressed throughout the brain without any good evidence of regional over-expression (MacDonald and Gusella, 1996). However, a common pathological feature has now emerged which will form the basis of further studies – the presence of **neuronal intranuclear inclusions** (**NII**) in both HD brains and transgenic mouse brains (see Mangiarini *et al.*, 1996; Ordway *et al.*, 1997). These abnormal, amyloid-like, structures are protein aggregates which are partially composed of expanded polyglutamines. In HD patients the distribution of NIIs corresponds with affected neuronal groups. Because huntingtin is normally cytoplasmic, the presence of NIIs in a subset of neurones therefore indicates that cell-specific, subcellular localisation of the mutant protein is an underlying mechanism of the disorder. Another of the hypothesised mechanisms of section 9.2.1 may also be involved, because it has been suggested that proteolytic cleavage of huntingtin is required for portions of this large protein to enter the nucleus (Figure 9.4).

Although the NIIs may reasonably be expected to interfere with neuronal function, what is their involvement in the neurodegenerative path-

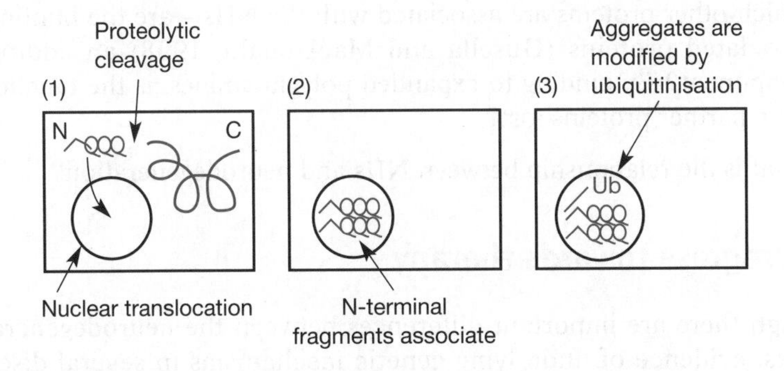

Figure 9.4 Hypothetical model for the generation of neuronal intranuclear inclusions in Huntington's disease (HD). It is proposed that the HD protein with an expanded polyglutamine (poly Q) region undergoes a conformational change, and is cleaved proteolytically to leave a small N-terminal, polyglutamine-containing fragment which is translocated to the nucleus. In the nucleus it associates with other similar fragments via a β-pleated sheet, the proteins are further modified by ubiquitination, and an intranuclear inclusion is formed. Modified from Ross, C.A. (1997) *Neuron*, **19**, 1147-1150.

way? Despite the presence of both a neurological phenotype, and NIIs in two of the described HD mouse models (Mangiarini *et al.*, 1996; Ordway *et al.*, 1997), there is no evidence of neuronal loss. It is therefore reasonable to question how accurately these models reflect the late-onset neurodegenerative diseases. On the other hand, a transgenic model of a related disorder, Machado–Joseph's disease, has provided evidence of expanded polyglutamine-dependent neuronal loss, apparently via an apoptotic pathway (Ikeda *et al.*, 1996).

9.4.3 Directions of further studies

The convergence of clinical findings and transgenic mouse studies with respect to the NIIs of HD (and other expanded polyglutamine neurodegenerative diseases) has fully justified the use of animal models in the search for the basis of HD. Further use of mouse models is now required for an understanding of disease progression. Current questions include:

- How do NIIs form ? Do expanded polyglutamine tracts oligomerise and form beta-sheets as predicted by Perutz (see MacDonald and Gusella, 1996). To what extent is a seeding mechanism involved (Lansbury, 1997)?
- Why do only certain cells in each disease develop the protein aggregates of NIIs – do the disease genes confer cell-specificity to the toxic polyglutamine expansions?
- What is the function of the NIIs – are they directly toxic, incidental, or beneficial – perhaps acting as a neuronal defence mechanism (Sisodia, 1998)?
- Which other proteins are associated with the NIIs – are the huntingtin-associated proteins (Gusella and MacDonald, 1998) an additional component? By binding to expanded polyglutamines, is the function of these partner proteins lost?
- What is the relationship between NIIs and neurodegeneration?

9.5 Progress towards therapy

Although there are important differences between the neurodegenerative diseases, evidence of underlying genetic mechanisms in several diseases indicates that gene therapy may provide the hope for effective treatment (see Further reading). Integration of this new and rapidly developing technology with other therapeutic strategies such as cell transplantation is also envisioned. The principle of genetic therapy for neurodegenerative diseases has been proven in one animal study in which over-expression of the cell death regulator Bcl-2 in a transgenic mouse model of Amytrophic lateral sclerosis has been shown to attenuate the magnitude of spinal cord motor neurone degeneration (Kostic *et al.*, 1997). To what extent is this 'proof-of-principle' relevant to human diseases? Evidence that neuronal loss *per se* may not underlie at least some of the major symptoms of neurodegenerative diseases (see Guest Box p.153) also suggests that considerably more basic research is required to identify appropriate therapeutic targets.

9.6 Summary

- A variety of distinct neurodegenerative diseases exhibit different underlying causes/genetic bases but may share a general 'proteinopathy' mechanism.
- Neurodegeneration is mediated by defined 'cell death' mechanisms.
- Alzheimer's disease (AD) is associated with mutations in various genes including amyloid precursor protein (*APP*), apolipoprotein E (*apoE*), presenilin 1 (*PS1*) and presenilin 2 (*PS2*).
- Learning and memory deficits (dementia) in AD may not simply reflect loss of neurones.
- Prion diseases such as BSE (mad cow disease) may be transmitted by an aberrant protein agent that modifies the structure and function of a similar endogenous protein.
- The genetic basis of Huntington's disease (HD), and other related diseases, is an expanded trinucleotide (CAG) repeat within a disease-specific gene.
- The molecular pathology of HD involves a 'gain-of-function' mechanism – translation of the expanded CAG repeat into a polyglutamine repeat may impair brain function through the deposition of neuronal intranuclear inclusions.

Further reading

ELLIS, R.E., YUAN, J. and HORVITZ, H.R. (1991) Mechanisms and functions of cell death. *Annual Review of Cell Biology*, **7**, 663–698.

A comprehensive review of cell death dating from 1991. This review precedes the recently fashionable work on neuronal cell death, but it can be useful to refer back to older reviews for alternative perspectives and insights.

Apoptosis – A special section. *Science*, **281**, 1301–1322 (1998).

DEARMOND, S.J., SANCHEZ, H., YEHIELY, F. *et al.* (1997) Selective neuronal targeting in prion disease. *Neurone*, **19**, 1337–1348.

Recent experimental data from Nobel prize winner Stanley Prusiner's group.

PRICE, D.L. and SISODIA, S.S. (1998) Mutant genes in familial Alzheimer's disease and transgenic models. *Annual Review of Neuroscience*, **21**, 479–505.

Review of animal models that considers their relevance to a variety of disease states including motor neurone disease, and prion diseases.

LEE, M.K., BORCHELT, D.R., WONG, P.C., SISODIA, S.S. and PRICE, D.L. (1996) Transgenic models of neurodegenerative diseases. *Current Opinion in Neurobiology*, **6**, 651–660.

WARRICK, J.M., PAULSON, H.L., GRAY-BOARD, G.L., BUI, Q.T., FISCHBECK, K.H., PITTMAN, R.N. and BONINI, N.M. (1998) Expanded polyglutamine protein forms nuclear inclusions and causes neural degeneration in *Drosophila*. *Cell*, **93**, 939–949.

The recent establishment of a polyglutamine repeat neurodegeneration model in fruit flies may provide fresh insights into neuropathological mechanisms.

VERMA, I.M. and SOMIA, N. (1997) Gene therapy – promises, problems and prospects. *Nature*, **389**, 239–242.

CHAPTER TEN

Behavioural disorders

Key points

- Schizophrenia
  - Neurochemical pathology
  - Developmental origin
  - Non-Mendelian inheritance
  - Identification of susceptibility loci
- Manic depression
  - Bipolar affective disorders
- Quantitative behavioural traits
  - QTL (quantitative trait loci analysis)
  - Animal models
- Stress

10.1 Introduction

Psychiatric illnesses such as schizophrenia and manic depression are disorders of behaviour that can be considered to be at the frontier of molecular neuroscience research. These pathologies are now within the grasp of molecular neuroscientists because recent inheritance studies (see section 10.2.3) have established that schizophrenia, for example, does have a strong genetic component and can therefore be investigated using molecular genetic approaches. However, unlike diseases such as Huntington's (see Chapter 9), it is clear that schizophrenia is not associated with a single aberrant gene, rather it is a **polygenic** disorder. This chapter will consider how the molecular basis of complex behavioural disorders can be approached.

Beyond the study of diseases such as schizophrenia, a 'final frontier' of molecular neuroscience is the analysis of **quantitative behavioural traits** (see section 10.4), such as mood and personality. Study of these human traits is justified from a medical standpoint because extremes in these individually variable behaviours represent very common illnesses such as anxiety disorders. This chapter will also describe current experimental approaches to these traits, a line of research which begins to question the molecular basis of consciousness.

General links to mental health Web sites including anxiety, sexual and eating disorders and schizophrenia can be found at: **http://www.mentalhealth.com.**

10.2 Schizophrenia

Schizophrenia is a complex, life-long disorder that is characterised by two types of clinical signs:

- Positive symptoms, e.g. hallucinations, delusions and bizarre behaviour.
- Negative symptoms, e.g. low motivation and expressed emotion.

Although 'schizophrenic' behaviour is commonly represented in popular culture (e.g. Hitchcock's 1960 movie, *Psycho,* **http://nextdch.mty.itesm.mx/~plopezg/Kaplan/mov/psychof/psycho.html**), in fact this condition remains one of the least understood of the major brain diseases. Genetic linkage has not been established (see section 10.2.5), and therefore the disease cannot yet be modelled in transgenic animals. Of course, a problem facing basic scientists in this field is that the perceptual deficits recognised in humans have no obvious equivalent in experimental animals.

Effective drug therapy for this psychiatric condition has been available for many years (see section 10.2.1), but there is no cure and repeated relapses and side effects during therapy lead to social and occupational dysfunction. The need for both a greater understanding of the biology of schizophrenia, and for the development of more effective treatments, is crucial because it has been estimated that schizophrenia afflicts 1% of the world's population. Aside from the healthcare cost, the social impact of schizophrenia is enormous – approximately 50% of people with schizophrenia attempt suicide, and at least one-third of homeless people are considered to be schizophrenic.

A basic introduction to this condition is available on the NIMH schizophrenia page: **http://www.nimh.nih.gov/publicat/schizo.htm#schiz3.**

10.2.1 Pathophysiology of schizophrenia

Characteristic pathological features of neurological disorders such as the amyloid plaques and neurofibrillary tangles of Alzheimer's disease (see section

9.3.1) provide important clues about the causes of disease. In the case of schizophrenia, there is macroscopic evidence of a reduction in cerebral cortex mass (see Egan and Weinberger, 1997), and microscopic evidence of changes in cortical neuronal organisation. Other aspects of pathology have been obtained from extensive neurochemical studies (see the following section).

Neurochemical changes in schizophrenia

Certain pharmacologically specific **psychotropic** drugs form an effective treatment for schizophrenia in some patients (see below), which is suggestive of a specific neurochemical deficit. Studies of brain neurochemistry, however, have indicated the presence of multiple different abnormalities (Figure 10.1).

- **Serotonin (5-HT).** The well-known hallucinogenic properties of 5-HT agonists such as **LSD,** and the anti-psychotic effects of some 5-HT antagonists have provided support for a 'serotonin hypothesis' of schizophrenia. The importance of changes in serotonergic function in schizophrenia has been questioned, although a recent review has hypothesised that regional differences in 5-HT activity may underlie both the positive and negative symptoms of schizophrenia (Breier, 1995).
- **Glutamate.** The perceptual distortions associated with drugs of abuse such as phencyclidine ('Angel Dust'), which is an NMDA receptor channel antagonist, have led to suggestions that glutamate function may

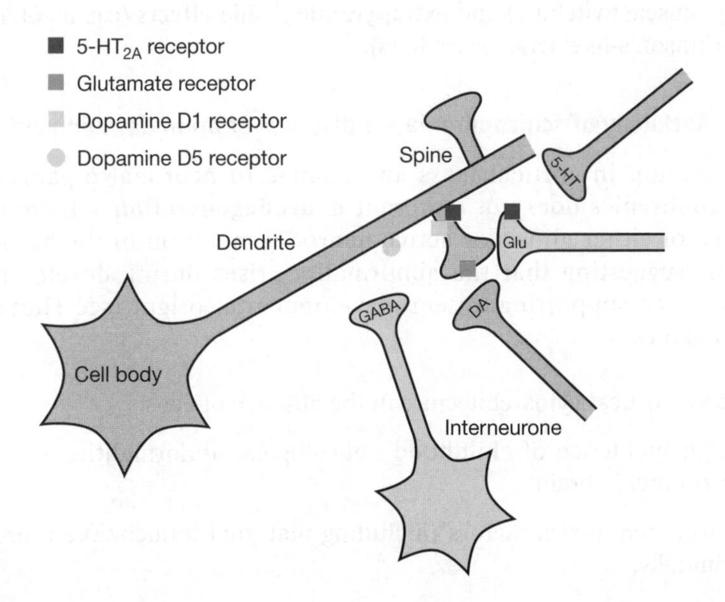

Figure 10.1 Simplified illustration of the synaptic organisation within the prefrontal cortex indicating the localisation of different receptors and neurotransmitters (shown in colour) which may be affected in schizophrenic brains. DA, dopamine; Glu, glutamate; 5-HT, serotonin. Modified from Nestler, E.J. (1997) *Nature*, **385**, 578–579.

be impaired in schizophrenics. This hypothesis has been supported by studies showing decreased expression of specific glutamate receptor transcripts in post-mortem brains (see Egan and Weinberger, 1997).

- **γ-Aminobutyric acid (GABA).** Changes in the levels of cortical GABA, GABA receptors, and numbers of GABA neurones indicate a reduction in GABA neurotransmission in schizophrenia, although GABA receptor subunit mRNA expression does not appear to be altered (see Egan and Weinberger, 1997).
- **Dopamine.** Despite the widespread and effective use of dopamine receptor antagonists (e.g. chlorpromazine, haloperidol) as controlling therapeutics for schizophrenia, the 'dopamine hypothesis' of schizophrenia is not supported by a well-defined neurochemical pathology. Thus, studies on dopamine receptor binding and gene expression in post-mortem brain samples are equivocal (see Egan and Weinberger, 1997). The use of modern imaging techniques (e.g. positron emission tomography – 'PET scan') to measure changes in dopamine receptor activity in the living brain appears to be a promising advance, provided that the results of initial studies (e.g. Okubo *et al.*, 1997) can be replicated. With respect to the general relevance of dopamine, it is also important to note that up to 30% of schizophrenics are resistant to treatment with dopaminergic drugs such as haloperidol and chlorpromazine. In addition, other patients are forced to stop this therapy because of major adverse effects, such as **tardive dyskinesia** (involuntary muscle twitching) and **extrapyramidal side effects** (e.g. involuntary 'Parkinson's-like' jerks or tremors).

10.2.2 Aetiology of schizophrenia – a disorder of brain development?

The reduction in cortical mass and change in neuronal organisation in schizophrenics does not represent **neurodegeneration** – there is no evidence of either gliosis or active neurodegeneration in the brains of patients, suggesting that the abnormality arises during development. The evidence supporting a neurodevelopmental origin (see Harrison, 1997) includes:

- altered cortical cytoarchitecture in the absence of gliosis;
- a high incidence of childhood neurological abnormalities – a 'pre-schizophrenic brain';
- environmental risk factors (including maternal influenza) act pre- or perinatally;
- cognitive deficits are present at the onset of symptoms and are not progressive.

10.2.3 Schizophrenia genetics

Family, twin and adoption studies have demonstrated that schizophrenia is predominantly a genetic disorder. For example, compared with the 1% risk in the general population, the risk to someone with an affected sibling and an affected parent is approximately 16%. Twins are said to be **concordant** if they both show a **discontinuous trait** – the concordance of schizophrenia in identical **(monozygotic)** twins is significantly higher (46–48%) than in non-identical **(dizygotic)** twins (4–14%) (see Karayiorgou and Gogos, 1997). The twin studies clearly show, however, that over half of monozygotic twins remain unaffected by the disorder, and because these twins essentially share all their genes, it is apparent that schizophrenia is *not* inherited in a simple **Mendelian** manner.

10.2.4 Non-Mendelian inheritance

Many common familial disorders such as schizophrenia exhibit a complex pattern of inheritance which is generally termed non-Mendelian. This complexity may arise in a variety of ways:

- **Polygenic inheritance** – multiple genes act additively to produce the phenotype. Inheriting one of the genes may produce a partial phenotype.
- **Heterogeneity** – a similar phenotype arises from different genotypes.
- **Incomplete penetrance** – the genotype is not always expressed as a phenotype. For example, an environmental co-factor may be required.
- **Anticipation** – the phenotype begins at an earlier age and increases in severity with succeeding generations. This is observed in Huntington's disease (see section 9.4.1) where it is related to expansion of a pathogenic trinucleotide repeat sequence. A trinucleotide repeat expansion of unknown pathogenic significance has been identified in a group of related schizophrenics (Sirugo *et al.*, 1997).
- **Imprinting** – gene expression and hence phenotype is dependent upon the sex of the parent transmitting the disease allele. This mode of inheritance is well characterised in a form of mental retardation termed the **fragile-X syndrome** – so-called because the X chromosome of transmitting males exhibits a so-called disease **premutation.**

10.2.5 Schizophrenia genes?

While other risk factors such as early environmental insults (see section 10.2.2) are almost certainly involved, it is apparent now that the onset of schizophrenia is dependent upon several **susceptibility loci.** These loci are thought to contain multiple genes that each make small contributions to the behavioural disorder.

GUEST BOX BY M.J. OWEN

Molecular psychiatry

Molecular genetics has been incredibly successful in identifying the genes responsible for a large number of single gene disorders. However, the current challenge is to adapt and refine these methods and apply them successfully to the really common causes of mortality and morbidity such as cardiovascular disease, diabetes, asthma, multiple sclerosis and the major psychiatric disorders. These common diseases are difficult for the geneticist because they show complex non-Mendelian patterns of inheritance. This is because they reflect the combined action of several, and in some cases possibly many, different genes which individually may be neither necessary nor sufficient to cause the disorder, as well as environmental factors. For the major psychiatric disorders such as schizophrenia and bipolar affective disorder, the problems do not stop here because in neither case can diagnosis be confirmed by any form of laboratory test or confirmed post mortem.

In spite of these difficulties, the potential rewards of finding genes for mental illness are great since these disorders are extremely common. Even in the era of 'community care', schizophrenia still accounts for more hospital beds than any other disorder! However, the brain is a very inaccessible organ and the diseases themselves highly complex, which means that we still know very little about the neurobiology of major psychiatric disorders. Consequently, although we have some reasonably effective treatments these still do not treat the 'core' of the disorders and their modes of action are largely unknown. Molecular genetics allows us potentially to identify the genes responsible for a condition simply by identifying chromosomal regions shared by affected members of the same family. This means that it is possible to identify disease genes without relying upon any knowledge of disease pathogenesis. Once disease genes have been identified this should offer a new way into understanding the pathophysiology of these disorders and hopefully lead to the development of rational and more effective therapies.

The greatest advances so far have been made in Alzheimer's disease. Advances in other common psychiatric disorders have as yet been more modest. However, a number of chromosomal regions likely to contain genes for schizophrenia and bipolar disorder have been identified, though more work in larger samples is still required. Also, it seems that variation in the 5HT2A and dopamine D3 receptor genes confers a risk, albeit small, of developing schizophrenia. Future challenges include investigating aspects of 'normal' behaviour such as personality which are under a degree of genetic control and which probably underlie a number of common psychiatric disorders such as mild depression and alcoholism. We also need to develop suitable animal models to allow the function of the genes identified to be investigated. Progress towards identifying the genes for mental illness

GUEST BOX Contd

may be hard and slow, but the potential benefits for patients and for the neurosciences are likely to be great.

CRADDOCK, N. and OWEN, M.J. (1996) Modern molecular approaches to psychiatric disease. *British Medical Bulletin*, **52**, 434–452.

Mike Owen completed a PhD in Behavioural Neuroscience and qualified in Medicine from the University of Birmingham. He then trained in Psychiatry at the Maudsley Hospital and in molecular genetics at St Mary's Hospital, London. In 1990, he moved to the University of Wales College of Medicine as Senior Lecturer in Neuropsychiatric Genetics. Since 1995 he has been Professor of Neuropsychiatric Genetics and from October 1998 will be Head of the Department of Psychological Medicine. He currently heads a research group attempting to locate and identify genes conferring susceptibility to psychiatric disorders, in particular schizophrenia, bipolar affective disorder (manic depression) and Alzheimer's disease.

The identification of schizophrenia susceptibility genes (see Guest Box above) can be attempted in many ways:

- Beyond **linkage analysis.** Two chromosomal loci are defined as being genetically linked when the probability of a recombination event occuring between them is <50%. In the analysis of disease linkage, disease loci are compared with known DNA marker loci, and statistical analysis **(lod scoring)** is used to determine whether the deviation from 50% is statistically significant. In non-Mendelian disease inheritance, however, a number of the assumptions of simple linkage analysis do not apply. New, more sophisticated linkage methods have been devised (see Karayiorgou and Gogos, 1997), and these have indicated the presence of schizophrenia susceptibility loci at chromosomal regions 22q, 8p and 6p. Currently, international collaborative studies of these loci are in progress, and schizophrenia-associated genes may soon be identified.
- Known **candidate genes.** Given the identified neurochemical pathologies (see section 10.2.1), and the availability of patient DNA sequence it is reasonable to adopt a candidate gene (informed guess) approach and search for abnormalities in neurotransmitter-related genes. A **polymorphism** in the dopamine D3 receptor gene for example, was shown to be associated with schizophrenia, but this finding has not been generally replicated (Egan and Weinberger, 1997). Future attempts at identifying candidate genes can now be informed by linkage analysis as described above.
- **SAGE** analysis. The possibility that differences in the level of expression of particular genes may underlie schizophrenia is currently being investigated by comparing SAGE profiles (Box 2.1) between normal and

schizophrenic brains. This approach is based (partly) on the premise that disease-related genetic variations may be located, not in the coding regions of particular genes, but in the regulatory sequences that control expression (see section 10.4).

10.2.6 Studies in experimental animals

What use are animal studies for the investigation of a complex psychiatric disorder? Neuropharmacological studies involving the administration of drugs such as phencyclidine (see section 10.2.1) have been widely employed as models of schizophrenia in the past, but do not provide any useful insights into either the molecular basis or clinical progression of the disease. Recently, one study has provided a potentially interesting model of disease progression in schizophrenia, showing that neonatal brain lesions in rats are associated with delayed behavioural effects (Lipska and Weinberger, 1995). Current **gene knockout** mouse models which exhibit 'schizophrenia'-like phenotypes (e.g. Dulawa *et al.*, 1997) have limited relevance to the molecular basis of the human condition because, as described above, schizophrenia is inherited as a polygenic disorder. In future studies it is likely that the functional interactions of multiple mutated genes will be investigated by crossing different knockout mouse lines.

10.3 Manic depression

Manic depression is another complex behavioural disorder which has recently been shown to have a strong genetic component (MacKinnon *et al.*, 1997). Clinically, manic depressive illness (MDI) is subclassified into different **bipolar disorders** (e.g. bipolar disorder type I, BPI), which relate to particular phenotypes. It is a very common illness, BPI alone representing 0.5–1.0% of the population, and current drug therapies such as lithium are inadequate. Concordance rates (see section 10.2.3) are reported to be 50–70% for monozygotic twins, and 13–30% for dizygotic twins. Other family and adoption studies have also provided evidence of inheritance. Genetic linkage studies have indicated MDI linkage to chromosome 18 (MacKinnon *et al.*, 1997) but, like schizophrenia, MDI exhibits non-Mendelian inheritance (see section 10.2.4) and the search is on for susceptibility genes which could be used in future animal model studies.

10.4 Quantitative behavioural traits

Genetic traits can be distinguished into two groups:

1 **Qualitative traits:** distinct extremes of phenotype such as deafness (see Guest Box p. 43) which are associated with a defined genotype. In genetic illnesses of this type a gene mutation results in altered protein function that is inconsistent with normal function.

2 **Quantitative traits:** normally distributed characteristics such as mood and intelligence, in which illnesses (e.g. anxiety disorder) form an extreme of a continuum that includes normality. The molecular basis of such illnesses is unlikely to be a specific gene mutation but rather subtle changes in gene expression, perhaps during critical periods in development.

The analysis of **quantitative behavioural traits** such as mood and personality is now an important area of molecular neuroscience research because it is hoped that this will lead to insights into the basis of certain psychiatric conditions, including some forms of depression and anxiety disorders.

10.4.1 Quantitative trait loci (QTL) analysis

QTL analysis (see Guest Box below) is an approach that is currently being used to determine the contribution of multiple different chromosomal loci to quantitative behavioural traits. A key question concerns the role animal models can play in mapping QTLs that are relevant to psychiatric conditions (Flint and Corley, 1996). These studies are still in their infancy, but the analysis of fearfulness in mice (rodent emotionality; Figure 10.2) has been found to exhibit a remarkably simple genetic basis (see Guest Box).

GUEST BOX BY J. FLINT

Quantitative trait loci (QTL) analysis of behavioural traits

Until recently, it was believed that the genetic architecture of behaviour would be so complex that it could be not be analysed with available tools. It was thought that there were many genes involved, each contributing a very small effect to the trait, thus rendering their individual detection extremely difficult. However, the application of quantitative trait analysis to animal behaviour, in which the contribution of multiple loci is determined, has been surprisingly successful. Results have been replicated and genes of relatively large effect (as much as 50% of the genetic variance of the trait) have been mapped: in crosses between inbred mice, three QTLs explain nearly 85% of the genetic variance of morphine preference and three loci account for virtually all the genetic variance of fearfulness ('emotionality'; see Flint *et al.*, 1995).

So animal studies show that QTL mapping of behaviour is possible. At present, many different behaviours are being mapped and the results are likely to be useful in providing a genetic validation for classification of behaviour. For instance, it has already been shown that genetic differences in fear conditioning map to the same place as differences in fearfulness, suggesting that the two traits have a common determinant.

Yet the encouraging results from animal studies may be misleading for the study of human behaviour. For one thing, the simplicity of the genetic architecture may reflect the simplicity of the genetic experiments: a cross between inbred animals has only two alleles segregating at each locus and the total amount of genetic

GUEST BOX Contd

variation is substantially less than in outbred populations. For another, we still do not know how frequently the the same QTLs will be found in different inbred strains, let alone in different species. Will QTL analysis work in humans? A QTL for reading disability has been found and replicated on chromosome 6p (Cardon *et al.*, 1994) so although the indications are hopeful, to date there have been no genome-wide QTL studies of behaviour.

The attraction of a QTL analysis of behaviour is that it promises to result in the cloning of genes. However, two important caveats need to be noted. The first is that cloning QTLs may prove far harder than expected. By definition, quantitative traits form a distribution which in most cases encompasses the normal, so it is hard to imagine that a QTL will consist of a genetic mutation that either inactivates a gene or leads to an abnormal gene product. It is more likely that the variant will result in small changes in gene expression. At present, we do not have the tools to recognise such genetic variants. The second caveat is that genetic variation may not occur in genes that are essential for the biology of a trait. For example, genes determining variation in an animal's sensitivity to pain, light or sound are likely to be involved in differences in behavioural responses to conditioning (which is often measured by pairing an electrical footshock to an auditory or visual cue). So it is conceivable that genetic investigation of conditioning may result in the isolation of genes involved in determining an animal's stimulus sensitivity, and will not teach us much about the biology of learning. However, numerous groups are now working in this field and it is likely that we will soon learn whether QTL mapping and cloning is worth the effort.

Flint, J., Corley, R., DeFries, J.C., Fulker, D.W., Gray, J.A., Miller, S. and Collins, A.C. (1995) A simple genetic basis for a complex psychological trait in laboratory mice. *Science*, **269**, 1432–1435.

Cardon, L.R., Smith, S.D., Fulker, D.W., Kimberling, W.J., Pennington, B.F. and DeFries, J.C. (1994) Quantitative trait locus for reading disability on chromosome 6. *Science*, **266**, 276–279.

Jonathon Flint trained in molecular biology in the Molecular Haematology Unit at Oxford under Professor David Weatherall, and in Psychiatry at the Maudsley and Bethlem Hospitals in London. He is currently a Wellcome Trust Senior Clinical Fellow at Oxford working on idiopathic mental retardation and the genetic basis of anxiety and depression.

Considerably more work will be required to characterise the relevant genes within QTLs, although progress should be enhanced by observations of specific behavioural deficits in gene knockout studies. An example is aggressive behaviour which has been shown to be affected in both nitric oxide synthase (Nelson *et al.*, 1995) and *tailless* (Monaghan *et al.*, 1997) gene knockout mice. Such profound mutations are almost certainly not directly relevant to the subtle allelic variations within quantitative behavioural traits, but these studies provide vital functional insights into the role of candidate genes.

10.5 Stress

Although not strictly speaking a behavioural disorder, stress is an active area of research in behavioural and molecular neuroscience. Interest in this field has arisen partly through the recognition of stress-related disorders that are associated with contemporary lifestyles. Much of the basic research on stress in recent years has centred on the potential actions of stress hormones such as **glucocorticoids** in causing brain cell death – the worrying scenario which emerges from these studies is that stress-related activation of glucocorticoid receptors in certain brain areas is associated with neuronal loss. The molecular mechanisms which undelie this process are currently being investigated; an excellent background to this field is provided in Robert Sapolsky's books (see Further reading). Stressful life events can also modulate the course of schizophrenia (see Harrison, 1997), and are also implicated in the development of drug addiction (see Box 7.1).

Figure 10.2 Differences in 'emotionality' (an animal correlate of human anxiety) between laboratory rodents can be assessed using an open-field test. When brightly lit, a circular arena is generally unpleasant for rats and mice, but there are measurable differences in open-field behaviour between different strains of animals, and these are genetically linked (see Box p. 171). This type of animal study forms one approach to the genetic analysis of quantitative human behavioural traits.

10.6 Summary

- Schizophrenia is a complex behavioural disorder characterised by changes in multiple neurochemical systems including dopamine and serotonin receptors.
- Brain structural changes in schizophrenia are not the result of neurodegeneration, rather a neurodevelopmental origin is indicated.
- Family, twin and adoption studies have demonstrated that schizophrenia is predominantly a genetic disorder, but it is apparent that schizophrenia is not inherited in a simple Mendelian manner.
- Non-Mendelian modes of inheritance may involve polygenic inheritance, genetic heterogeneity, incomplete penetrance, anticipation and imprinting.
- Schizophrenia susceptibility loci have been found on three human chromosomes.
- Schizophrenia appears to involve many genes, each with small effects.
- Manic depression is inherited in a similar non-Mendelian manner.
- The genetic basis of quantitative behavioural traits can be approached in animal models using QTL analysis.
- Behavioural stress can affect both neuronal survival and psychiatric disease.

Further reading

HARMAN, D.S. and CIVELLI, O. (1997) Dopamine receptor diversity: molecular and pharmacological perspectives. *Progress in Drug Research*, **48**, 173–194.

A recent review of dopamine receptors authored by Oliver Civelli, one of the pioneers of dopamine receptor cloning.

MCGUE, M. and BOUCHARD, T.J. (1998) Genetic and environmental influences on human behavioural differences. *Annual Review of Neuroscience*, **21**, 1–24.

A recent review of the 'nature–nurture' debate.

RISCH, N. and MERIKANGAS, K. (1996) The future of genetic studies of complex human diseases. *Science*, **273**, 1516–1517.

SAPOLSKY, R.M. (1994) *Why zebras don't get ulcers: a guide to stress, stress related diseases, and coping*. W.H. Freeman, New York.

Some light reading from one of the leading researchers in the field of animal stress.

Glossary

alternative splicing Different mRNAs may be formed through differential splicing of an RNA precursor.

allogeneic Within a species.

annealing The association (hybridisation) of two complementary strands of DNA into a duplex molecule.

antisense RNA which is complementary to an mRNA sequence.

apoptosis Programmed cell death (cf. necrosis).

autosome Any chromosome other than the sex chromosomes.

BAC *See* bacterial artificial chromosome.

bacterial artificial chromosome Synthetic cloning vector which can accommodate very large fragments of DNA.

band-shift [gel-shift; electrophoretic mobility shift assay (EMSA)] A laboratory technique used to detect the binding of proteins to specific DNA sequences.

bioinformatics The compilation of biological information (DNA sequences, protein structures) into rational databases through the application of computer and statistical techniques.

biolistics Laboratory techniques which involve firing DNA into cells/tissues from a 'gene gun'.

candidate gene A known gene which is implicated in a particular function/dysfunction.

cDNA See complementary DNA.

cDNA library A collection of cDNA molecules which represents a particular mRNA population.

chemical mutagenesis A laboratory technique used to cause random mutations in DNA.

chimera An animal composed of elements derived from genetically distinct individuals. Can also refer to synthetic genes/proteins which have been engineered to contain sequences from two or more genes/proteins.

circadian rhythm A rhythm of approximately 24-hour duration which is maintained in a constant environment.

***cis*-acting element** A DNA regulatory element which controls activity of a gene on the same chromosome. Can also refer to an RNA element which controls transcript activity.

clock gene A gene which encodes an inherent component of a circadian clock mechanism.

clock-controlled gene A gene which exhibits a rhythmic pattern of expression in circadian clock cells but is not an inherent component of a circadian clock mechanism.

complementary DNA A reverse complement copy of a mRNA sequence synthesised by the enzyme reverse transcriptase.

concordant Both members of a twin pair exhibit a trait.

congenic Two strains that are genetically identical except for a difference at a single locus. In practice, strains differ by a chromosomal segment rather than in a single gene. Congenic strains are produced by backcrossing into an inbred strain background. A strain developed by repeated backcrossing should be regarded as congenic when a minimum of 10 backcross generations to the background strain have been made, counting the first hybrid or F_1 generation as generation 1.

contig Overlapping contiguous cloned DNA fragments.

cosmid Synthetic cloning vector which can accommodate large fragments of DNA.

degenerate Refers to a collection of variant DNA oligonucleotides which have been designed to accommodate the degeneracy of the genetic code, and thereby represent all (or most) sequences which code for a particular amino acid sequence.

denaturation Conversion from the double-stranded into the single-stranded form of DNA (normally accomplished by boiling).

discontinuous trait A trait (characteristic) which exhibits variations that fall into two or more discrete classes.

dominant (mutation) A phenotypic trait expresssed in heterozygotes.

embryonal stem cell Pluripotent cells in the blastocyst which give rise to all the differentiated cells of the embryo.

epitope An antigenic determinant; an amino acid sequence or molecule that elicits an antibody response in an immunised animal.

ES cell *See* embryonal stem cell.

EST *See* expressed sequence tag.

excitotoxicity Toxic action mediated through an excess of excitatory neurotransmitter.

expressed sequence tag A randomly cloned cDNA fragment of an RNA.

expression library A collection of cDNA molecules in a lambda phage vector that expresses the DNA as proteins which can be detected with antibodies (immunoscreening).

familial Inherited (within families). Implies a genetic basis of a trait, although epigenetic factors should not be disregarded.

forced evolution A laboratory technique in which a gene sequence is randomly rearranged in order to obtain proteins with modified (e.g. enhanced) functional activity.

forward genetics From phenotype to genotype; (cf. reverse genetics). Refers to an experimental approach that seeks to discover the gene underlying a particular functional/morphological characteristic.

frameshift An alteration (mutation) in DNA or RNA sequence which results in a downstream change in the coding potential of the resultant mRNA.

full-length Refers to the size of a cDNA molecule. It is a cDNA which is a complete copy of the corresponding mRNA.

gain-of-function Refers to an experimental strategy or disease mutation which does not cause the effect/pathology through interfering with the normal function of a gene – rather a deviant property/function is generated by the strategy/mutation.

genetic anticipation Earlier ages of disease onset (and severity) in successive generations, e.g. in Huntington's disease.

genome Total genetic information carried by an organism.

genomic library A collection of fragments of chromosomal DNA in a vector representing most of the genome.

genotype The genetic constitution of an organism.

germ cell A cell that gives rise to reproductive cells.

germline The lineage of cells that gives rise to the germ cells and transmits genetic information to subsequent generations.

hemizygous A gene present in only one dose, e.g. as in a transgenic line in which a transgene is inserted (perhaps in multiple copies) in a particular chromosomal location. Note that matings between transgenic individuals will give rise to some homozygous offspring (if homozygotes are viable).

homologous recombination Genetic exchange between similar (if not identical) DNA sequences.

homology screening A library (e.g. cDNA library) screen in which a probe derived from a known gene is used to detect closely related genes.

homozygous Having an identical allele of a gene in a pair of homologous chromosomes.

heterozygous Having different (e.g. one normal and one mutant) alleles in a pair of homologous chromosomes.

hyperphosphorylated Refers to a protein which exhibits an (abnormal) abundance of covalently linked phosphate groups.

immediate-early gene Gene (e.g. c-*fos*) whose expression is low or undetectable in quiescent cells, but whose transcription is activated within minutes after extracellular stimulation, such as the addition of a growth factor.

imprinting A mechanism of inheritance in which DNA is modified during gametogenesis so affecting gene expression.

insertional mutation A gene sequence is interrupted following chromosomal insertion of a foreign (e.g. transgene) DNA sequence.

***in situ* hybridisation** A laboratory technique used to localise RNA or DNA molecules in cells/tissues.

instability (of DNA) Certain simple repeat sequences in the genome (e.g. CAGCAGCAG, etc.) have an inherent instability and can exhibit expansion across generations. The mechanism of expansion is undefined but may involve 'slippage' of DNA polymerase during replication.

junk Refers to the portion of chromosomal DNA (over 90% of the mammalian genome) composed of repetitive elements that have no known function either as sequences that code for proteins or regulate gene expression.

knock-down Common term for a laboratory technique in which levels of a specific transcript are reduced (e.g. using a synthetic antisense sequence).

knock-in Common term for a laboratory technique in which a genomic DNA sequence is replaced in a targeted fashion such that a specific mutation is introduced into a gene.

knock-out Common term for a laboratory technique in which a genomic DNA sequence is replaced in a targeted manner such that a specific gene is inactivated.

Kozak consensus sequence The sequence surrounding the translation initiation codon (ATG) which is highly conserved and promotes ATG selection in preference to non-consensus sequences. Named after Marylyn Kozak.

linkage Genes are described as linked when found to be inherited more frequently than expected by chance.

locus The location of a particular gene on a chromosome.

lod score Logarithm of the odds score for the likelihood of two genetic loci being within a measurable distance of one another.

long-term potentiation A persistent strengthening of synaptic transmission based on past activity.

loss-of-function Refers to an experimental strategy or disease mutation which impairs either the normal expression or function of a gene.

LTP *See* long-term potentiation.

mRNA *See* messenger RNA.

melting temperature (T_m) The temperature at which strands of duplex DNA denature.

Mendelian Refers to inheritance/genetics that follows a simple pattern as initially defined by Gregor Mendel.

messenger RNA (mRNA) An RNA molecule transcribed from a gene sequence and translated to yield a protein.

molecular misreading Aberrant RNA sequences are transcribed from a normal DNA sequence.

necrosis Cell death that is not programmed (cf. apoptosis).

neurodegeneration Functional deterioration and eventual loss of neurones.

non-Mendelian Refers to inheritance/genetics which does not follow a simple Mendelian pattern.

Northern blotting A laboratory technique used for the visualisation of mRNA molecules.

nuclear run-on analysis/assay A laboratory technique which directly measures gene transcription.

nuclear transfer A laboratory technique in which the nucleus of one cell (type) is transferred to an enucleated cell.

open reading frame mRNA sequence which codes for a continuous stretch of amino acids, not interrupted by STOP codons.

ORF *See* open reading frame.

orphans Refers to cloned genes, for example receptor genes, which have no known ligand/interacting protein.

over-expressed Refers to an experimental paradigm in which a gene is expressed at supraphysiological levels in order to investigate functional consequences.

PCR *See* polymerase chain reaction.

penetrance (incomplete) The extent to which a genotype is expressed as a phenotype.

phenotype The observable characteristics of an organism.

phosphorylation Covalent linkage of a phosphate group to a molecule, promoted by protein kinases and reversed by protein phosphatases.

poly (A) addition site Sequence (AAUAAA) which mediates cleavage of the primary transcript and addition of the terminal poly (A) tail on mRNA molecules. Normally found 10–30 nucleotides upstream of the poly (A) tail.

polygenic Determined by multiple genes, each with a small but additive effect.

polymerase chain reaction A laboratory technique for the exponential synthesis of specific DNA fragments.

polymorphic gene A gene which exhibits more than one relatively common allele.

polymorphism Common discontinuous genetic variation.

positional cloning A strategy of gene cloning based on the position of the gene in a genetic/physical map. Unlike other cloning strategies, sequence information is not required.

precursor RNA (pre-mRNA) The primary transcript and processing intermediates that yield mRNA.

primer extension A laboratory technique used to determine the 5' termini of mRNAs.

prion Proteinaceous transmissable agent of neurodegenerative disease (probably).

promoter-reporter gene Commonly used synthetic DNA construct designed to demonstrate ('report') the activity of the promoter. Examples of reporter genes are chloramphenicol acetyltransferase (CAT), firefly luciferase, and jelly-fish green fluorescent protein (GFP).

QTL *See* quantitative trait loci analysis.

qualitative trait A distinct extreme of a phenotype such as deafness.

quantitative trait A normally distributed characteristic such as mood in which illnesses (e.g. anxiety disorder) form an extreme of a continuum that includes normality.

quantitative trait loci (QTL) analysis Determination of the contribution of different loci (*See* locus) to a quantitative trait.

RACE *See* rapid amplification of cDNA ends.

rapid amplification of cDNA ends A laboratory technique used to extend the known sequence of a cDNA.

recessive (mutation) A phenotypic trait expressed only in homozygotes.

recombination frequency The frequency at which chromosomal markers are inherited together. Provides an estimate of physical distance between the markers.

redundancy A phenomenon which is revealed when alternative genes, possibly of the same family, take over the function of knocked-out genes.

repeat DNA molecules exhibit repeats of simple combinations of nucleotides (e.g. a trinucleotide repeat such as CAGCAGCAG, etc.).

replication (of DNA) The copying of a DNA molecule.

reporter (gene) A gene whose expression can be monitored. Certain reporter genes are commonly selected for experiments because the gene products are not normally expressed in the experimental cell type.

restriction enzymes Enzymes which cleave ('cut') DNA molecules at defined sequences ('cutting sites').

retrovirus A type of animal virus containing an RNA genome that replicates in infected cells by making an complementary DNA copy of the RNA.

reverse genetics From genotype to phenotype (cf. forward genetics) Refers to an experimental approach such as transgenesis in which the function of a (perhaps novel) gene sequence is investigated.

reverse transcriptase An enzyme that synthesises complementary DNA from a single-stranded RNA template.

reverse transcriptase–polymerase chain reaction A laboratory technique for the exponential synthesis of DNA molecules starting from a specific mRNA template.

restriction fragment length polymorphisms Variations in DNA sequences that give rise to restriction enzyme cleavage sites that are present in some individuals, but not others.

RFLPs *See* restriction fragment length polymorphisms.

RNA editing A nuclear, post-transcriptional, mechanism in which the sequence of RNA is subtly altered.

RNAse protection A laboratory technique used to analyse mRNA (level and structure).

RT–PCR *See* reverse transcriptase polymerase chain reaction.

SAGE *See* serial analysis of gene expression.

sequence tagged site A DNA sequence (unique in the genome) which may be amplified using PCR and used to identify DNA clones.

serial analysis of gene expression A laboratory technique which provides a comprehensive (i.e. genome-wide), and quantitative analysis of mRNAs.

shotgun sequencing Random sequencing of DNA clones.

silent mutation Genetic mutation in which a codon sequence is altered, but because of the degeneracy of the genetic code, the encoded amino acid is unchanged.

simple sequence length polymorphism Genetic markers used to map chromosomes which correspond to variations in the lengths of simple sequence repeats.

somatic cell Any cell other than gametes and germ cells.

Southern blotting A laboratory technique used to visualise specific DNA molecules.

sporadic No apparent inherited component because the disease does not run in families, and therefore factors such as environment may be the determining factor, perhaps acting in combination with a genetic predisposing factor.

SSLP *See* simple sequence length polymorphism.

STOP codons Codons which signal termination of translation (UAA,UAG, UGA).

stringency Refers to the washing conditions used to remove probes which are hybridised to DNA sequences.

STS *See* sequence tagged site.

***trans*-acting factor** A diffusible protein (e.g. transcription factor) which controls gene activity through an interaction with a *cis*-acting element.

transcription factor General term for proteins (not RNA polymerase) required to initiate or regulate transcription.

transcription unit A DNA sequence, including regulatory sequences, that is transcribed into a primary transcript.

transgene A foreign gene sequence which has been introduced into a cellular organism by transgenesis.

transgenic Any cell or organism into which novel DNA has been introduced.

transgenic rescue The correction of a genetic defect by transgenesis.

transgenesis The application of transgenic technology – the introduction of new genes into living cells or organisms.

vector A synthetic DNA molecule (e.g. a plasmid) into which foreign DNA is inserted for cloning or expression.

Western blotting A laboratory technique used to visualise proteins.

wild-type The common ('normal') genotype/phenotype in a population of organisms.

xenogeneic Different species.

YAC *See* yeast artificial chromosome.

yeast artificial chromosome Synthetic cloning vector which can accommodate very large fragments of DNA.

yeast two-hybrid assay A laboratory technique which uses a protein as 'bait' to detect interacting proteins expressed from a cDNA library.

References

ABDALLAH, B., HASSAN, A., BENOIST, C., GOULA, D., BEHR, J.P. and DEMENEIX, B.A. (1996) A powerful non-viral vector for in vivo gene transfer into the adult mammalian brain: polyethylenimine. *Human Gene Therapy,* **20**, 1947–1954.

ADAMS, M.D., KERLAVAGE, A.R., FIELDS, C. and VENTER, J.C. (1993) 3,400 new expressed sequence tags identify diversity of transcripts in human brain. *Nature Genetics,* **4**, 256–267.

AGUZZI, A. and WEISSMAN, C. (1997) Prion research: the next frontiers. *Nature,* **389**, 795–798

ANTOCH, M.P., SONG, E.J., CHANG, A.M., VITATERNA, M.H., ZHAO, Y., WILSBACHER, L.D., SANGORAM, A.M., KING, D.P., PINTO, L.H. and TAKAHASHI, J.S. (1997) Functional identification of the mouse circadian *Clock* gene by transgenic BAC rescue. *Cell,* **89**, 655–667.

BEYREUTHER, K. and MASTERS, C.L. (1996) Tangle disentanglement. *Nature,* **383**, 476–477.

BEHRINGER, R.R., MATHEWS, L.S., PALMITER, R.D. and BRINSTER, R.L. (1988) Dwarf mice produced by genetic ablation of growth hormone-expressing cells. *Genes and Development,* **2**, 453–461.

BILANG-BLEUEL, A., REVAH, F., COLIN, P., LOCQUET, I., ROBERT, J.J., MALLET, J. and HORELLOU, P. (1997) Intrastriatal injection of an adenoviral vector expressing glial-cell-line-derived neurotrophic factor prevents dopaminergic neuron degeneration and behavioral impairment in a rat model of Parkinson disease. *Proceedings of the National Academy of Sciences USA,* **94**, 8818–8823.

BLISS, T.V. and COLLINGRIDGE, G.L. (1993) A synaptic model of memory: long term potentiation in the hippocampus. *Nature,* **361**, 31–39.

Bloom, F.E. (1996) An internet review: the compleat neuroscientist scours the World Wide Web. *Science,* **274,** 1104–1108.

Borchelt, D.R., Ratovitski, T., van Lare, J., Lee, M.K., Gonzales, V., Jenkins, N.A., Copeland, N.G., Price, D.L. and Sisodia, S.S. (1997) Accelerated amyloid deposition in the brains of transgenic mice coexpressing mutant presenilin 1 and amyloid precursor proteins. *Neuron,* **19,** 939–945.

Borrelli, E., Heyman, R.A., Arias, C., Sawchenko, P.E. and Evans, R.M. (1989) Transgenic mice with inducible dwarfism. *Nature,* **339**, 538–541.

Bourtchuladze, R., Frenguelli, B., Blendy, J., Cioffi, D., Schutz, G. and Silva, A.J. (1994) Deficient long-term memory in mice with a targeted mutation of the cAMP-responsive element-binding protein. *Cell,* **79,** 59–68.

Breier, A. (1995) Serotonin, schizophrenia and antipsychotic drug action. *Schizophrenia Research,* **14,** 187–202.

Brinster, R.L. and Zimmermann, J.W. (1994) Spermatogenesis following male germ-cell transplantation. *Proceedings of the National Academy of Sciences USA,* **91,** 11298–11302.

Brook, F.A. and Gardner, R.L. (1997) The origin and efficient derivation of embryonic stem cells in the mouse. *Proceedings of the National Academy of Sciences USA,* **94,** 5709–5712.

Brown, S.D.M. and Peters, J. (1996) Combining mutagenesis and genomics in the mouse – closing the phenotype gap. *Trends in Genetics,* **12,** 433–435.

Bruce, M.E., Will, R.G., Ironside, J.W., McConnell, I., Drummond, D., Suttie, A., McCardle, L., Chree, A., Hope, J., Birkett, C., Cousens, S., Fraser, H. and Bostock, C.J. (1997) Transmissions to mice indicate that 'new variant' CJD is caused by the BSE agent. *Nature,* **389,** 498–501.

Brussaard, A.B., Kits, K.S., Baker, R.E., Willems, W.P.A., Leyting-Vermeulen, J.W., Voorn, P., Smit, A.B., Bicknell, R.J. and Herbison, A.E. (1997) Plasticity in fast synaptic inhibition of adult oxytocin neurons caused by a switch in $GABA_A$ subunit expression. *Neuron,* **19,** 1103–1114.

Caine, S.B. (1998) Cocaine abuse: hard knocks for the dopamine hypothesis. *Nature Neuroscience,* **1,** 90–92.

Cao, L., Zheng, Z.C., Zhao, Y.C., Jiang, Z.H., Liu, Z.G., Chen, S.D., Zhou, C.F. and Liu, X.Y. (1995) Gene therapy of Parkinson disease model rat by direct injection of plasmid DNA–lipofectin complex. *Human Gene Therapy,* **6,** 1497–1501.

Chen, J., Kelz, M.B., Hope, B.T., Nakabeppu, Y. and Nestler, E.J. (1997) Chronic Fos-related antigens: stable variants of ΔFosB induced in brain by chronic treatments. *Journal of Neuroscience,* **17,** 4933–4931.

CHOI-LUNDBERG, D.L., LIN, Q., CHANG, Y.N., CHIANG, Y.L., HAY, C.M., MOHAJERI, H., DAVIDSON, B.L. and BOHN, M.C. (1997) Dopaminergic neurons protected from degeneration by GDNF gene therapy. *Science,* **275,** 838–841.

CRAMERI, A., WHITEHORN, E.A., TATE, E. and STEMMER, W.P.C. (1995) Improved green fluorescent protein by molecular evolution using DNA shuffling. *Nature Biotechnology,* **14,** 315–319.

DE STROOPER, B., SAFTIG, P., CRAESSAERTS, K., VANDERSTICHELE, H., GUHDE, G., ANNAERT, W., VON FIGURA, K. and VAN LEUVEN, F. (1998) Deficiency of pre-senilin-1 inhibits the normal cleavage of amyloid precursor protein. *Nature,* **391,** 387–389.

DIAMOND, M.I., MINER, J.N., YOSHINAGA, S.K. and YAMAMOTO, K.R. (1990) Transcription factor interactions: selectors of positive and negative regulation from a single DNA element. *Science,* **249,** 1266–1272.

DI MARZO, V., FONTANA, A., CADAS, H., SCHINELLI, S., CIMINO, G., SCHWARTZ, J.C. and PIOMELLI, D. (1994) Formation and inactivation of endogenous cannabinoid anandamide in central neurons. *Nature,* **372,** 686–691.

DRAGANOW, M. (1996) A role for immediate-early transcription factors in learning and memory. *Behaviour Genetics* **26,** 293–299.

DULAWA, S.C., HEN, R., SCEARCE-LEVIE, K. and GEYER, M.A. (1997) Serotonin1B receptor modulation of startle reactivity, habituation, and prepulse inhibition in wild-type and serotonin 1B knockout mice. *Psychopharmacology,* **132,** 125–134.

DUNLAP, J. (1998) An end in the beginning. *Science,* **280,** 1548–1549.

DURING, M.J., NAEGELE, J.R., O'MALLEY, K.L. and GELLER, A.I. (1994) Long-term behavioral recovery in Parkinsonian rats by an HSV vector expressing tyrosine hydroxylase. *Science,* **266,** 1399–1403.

EBISAWA, T., KARNE, S., LERNER, M.R. and REPPERT, S.M. (1994) Expression cloning of a high affinity melatonin receptor from *Xenopus* dermal melanophores. *Proceedings of the National Academy of Sciences, USA,* **91,** 6133–6137.

EGAN, M.F. and WEINBERGER, D.R. (1997) Neurobiology of schizophrenia. *Current Opinion in Neurobiology,* **7,** 701–707.

FENG, Y., ZHANG, F., LOKEY, L.K., CHASTAIN, J.L., LAKKIS, L., EBERHART, D. and WARREN, S.T. (1995) Translational suppression by trinucleotide repeat expansion at FMR1. *Science,* **268,** 731–734.

FLINT, J. and CORLEY, R. (1996) Do animal models have a place in the genetic analysis of quantitative human behavioural traits. *Journal of Molecular Medicine,* **74,** 515–521.

GEDDES, B.J., HARDING, T.C., LIGHTMAN, S.L. and UNEY, J.B. (1997) Long-term gene therapy in the CNS: reversal of hypothalamic diabetes insipidus in the Brattleboro rat by using an adenovirus expressing arginine vasopressin. *Nature Medicine,* **3,** 1402–1404.

GEERTZ, C. (1973) *The Interpretation of Cultures: Selected Essays.* Basic Books, New York, pp. 28–29.

GEKAKIS, N., STAKNIS, D., NGUYEN, H.B., DAVIS, F.C., WILSBACHER, L.D., KING, D.P., TAKAHASHI, J.S. and WEITZ, C.J. (1998) Role of the CLOCK protein in the mammalian circadian mechanism. *Science,* **280,** 1564–1569.

GRABOWSKI, P.J. (1998) Splicing regulation in neurons: tinkering with cell-specific control. *Cell,* **92,** 709–712.

GRANT, S.G., KARL, K.A., KIEBLER, M.A. and KANDEL, E.R. (1995) Focal adhesion kinase in the brain: novel subcellular localization and specific regulation by Fyn tyrosine kinase in mutant mice. *Genes and Development,* **9,** 1909–1921.

GREEN, T., HEINEMANN, S.F. and GUSELLA, J.F. (1998) Molecular neurobiology and genetics: investigation of neural function and dysfunction. *Neuron,* **20,** 427–444.

GU, H., MARTH, J.D., ORBAN, P.C., MOSSMANN, H. and RAJEWSKY, K. (1994) Deletion of DNA polymerase β gene segment in T cells using cell type-specific gene targeting. *Science,* **263,** 103–106.

GUSELLA, J.F. and MACDONALD, M.E. (1998) Huntingtin: a single bait hooks many species. *Current Opinion in Neurobiology* **8,** 425–430.

HANNAS-DJEBBARA, Z., DIDIER-BAZS, M., SACCHETTONI, S., PROD'HON, C., JOUVET, M., BELIN, M.F. and JACQUEMONT, B. (1997) Transgene expression of plasmid DNAs directed by viral or neural promoters in the rat brain. *Molecular Brain Research,* **46,** 91–99.

HARRISON, P.J. (1997) Schizophrenia: a disorder of neurodevelopment. *Current Opinion in Neurobiology,* **7,** 285–289.

HEBB, D.O. (1949) *The Organisation Of Behaviour.* Wiley, New York.

HERDEGAN, T. and ZIMMERMANN, M. (1995) Immediate early genes (IEGs) encoding inducible transcription factors (ITFs) and neuropeptides in the nervous system: functional network for long-term plasticity and pain. *Progress in Brain Research,* **104,** 299–321.

HERMAN, J.P. (1995) *In situ* hybridization analysis of vasopressin gene transcripition in the paraventricular and supraoptic nuclei of the rat: regulation by stress and glucocorticoids. *Journal of Comparative Neurology,* **363,** 15–27.

Huntington's Disease Collaborative Research Group (1993) A novel gene containing a trinucleotide repeat that is expanded and unstable on Huntington's disease chromosomes. *Cell,* **72,** 971–983.

IKEDA, H., YAMAGUCHI, M., SUGAI, S., AZE, Y., NARUMIYA, S. and KAKIZUKA, A. (1996) Expanded polyglutamine in the Machado–Joseph disease protein induces cell death *in vitro* and *in vivo*. *Nature Genetics,* **13,** 196–202.

IMPEY, S., MARK, M., VILLACRES, E.C., POSER, S., CHAVKIN, C. and STORM, D.R. (1996) Induction of CRE-mediated gene expression by stimuli that generate long-lasting LTP in CA1 of the hippocampus. *Neuron,* **16,** 973–982.

KANTOR, D.B., LANZREIN, M., STARY, S.J., SANDOVAL, G.M., SMITH, W.B., SULLIVAN, B.M., DAVIDSON, N. and SCHUMAN, E.M. (1996) A role for endothelial NO synthase in LTP revealed by adenovirus-mediated inhibition and rescue. *Science,* **274,** 1744–1748.

KAPLITT, M.G., LEONE, P., SAMULSKI, R.J., XIAO, X., PFAFF, D.W., O'MALLEY, K.L. and DURING, M.J. (1994) Long-term gene expression and phenotypic correction using adeno-associated virus vectors in the mammalian brain. *Nature Genetics,* **8,** 148–154.

KARAYIORGOU, M. and GOGOS, J.A. (1997) A turning point in schizophrenia genetics. *Neuron,* **19,** 967–979.

KASOF, G.M., MANDELZYS, A., MAIKA, S.D., HAMMER, R.E., CURRAN, T. and MORGAN, J.I. (1995) Kainic-acid induced neuronal cell death is associated with DNA damage and a unique immediate-early gene response in c-*fos-lacZ* transgenic rats. *Journal of Neuroscience,* **15,** 4238–4249.

KING, D.P., ZHAO, Y., SANGORAM, A.M., WILSBACHER, L.D., TANAKA, M., ANTOCH, M.P., STEEVES, T.D.L., VITATERNA, M.H., KORNHAUSER, J.M., LOWREY, P.L., TUREK, F.W. and TAKAHASHI, J.S. (1997) Positional cloning of the mouse circadian *Clock* gene. *Cell,* **89,** 641–653.

KISTNER, A., GOSSEN, M., ZIMMERMANN, F., JERECIC, J., ULLMER, C., LUBBERT, H. and BUJARD, H. (1996) Doxycycline-mediated quantitative and tissue-specific control of gene expression in transgenic mice. *Proceedings of the National Academy of Sciences,* USA, **93,** 10933–10938.

KOJIMA, N., WANG, J., MANSUY, I.M., GRANT, S.G., MAYFORD, M. and KANDEL, E.R. (1997) Rescuing impairment of long-term potentiation in *fyn*-deficient mice by introducing Fyn transgene. *Proceedings of the National Academy of Sciences,* USA, **94,** 4761–4765.

KOSTIC, V., JACKSON-LEWIS, V., DE BILBAO, F., DUBOIS-DAUPHIN, M. and PRZEDBORSKI, S. (1997) Bcl-2: prolonging life in a transgenic mouse model of familial amyotrophic lateral sclerosis. *Science,* **277,** 559–563.

KOZAK, M. (1996) Interpreting DNA sequences: some insights from studies on translation. *Mammalian Genome,* **7,** 563–574.

KRULEWSKI, T.F., NEUMANN, P.E. and GORDON, J.W. (1989) Insertional mutation in a transgenic mouse allelic with Purkinje cell degeneration. *Proceedings of the National Academy of Sciences, USA,* **86,** 3709–3712.

KUHN, R., SCHWENK, F., AGUET, M. and RAJEWSKY, K. (1995) Inducible gene targeting in mice. *Science,* **269,** 1427–1429.

LANSBURY, P.T. (1997) Structural neurology: are seeds at the root of neuronal degeneration? *Neuron,* **19,** 1151–1154.

LIN, C-L.G., BRISTOL, L.A., JIN, L., DYKES-HOBERG, M., CRAWFORD, T., CLAWSON, L. and ROTHSTEIN, J.D. (1998) Aberrant RNA processing in a neurodegenerative disease: the cause for absent EAAT2, a glutamate transporter, in amyotrophic lateral sclerosis. *Neuron,* **20,** 589–602.

LIPSKA, B.K. and WEINBERGER, D.R. (1995) Genetic variation in vulnerability to the behavioural effects of neonatal hippocampal damage in rats. *Proceedings of the National Academy of Sciences, USA,* **92,** 8906–8910.

LIU, C., WEAVER, D.R., STROGATZ, S.H. and REPPERT, S.M. (1997) Cellular construction of a circadian clock: period determination in the suprachiasmatic nuclei. *Cell,* **91,** 855–860.

LUCKMAN, S.M., DYE, S. and COX, H.J. (1996) Induction of members of the Fos/Jun family of immediate-early genes in identified hypothalamic neurons: *in vivo* evidence for differential regulation. *Neuroscience,* **73,** 473–485.

MACDONALD, M.E. and GUSELLA, J.F. (1996) Huntington's disease: translating a CAG repeat into a pathogenic mechanism. *Current Opinion in Neurobiology,* **6,** 638–643.

MACKINNON, D.F., REDFIELD-JAMISON, K.R. and DEPAULO, J.R. (1997) Genetics of manic depressive illness. *Annual Review of Neuroscience* **20,** 355–373.

MANDEL, R.J., SPRATT, S.K., SNYDER, R.O. and LEFF, S.E. (1997) Midbrain injection of recombinant adeno-associated virus encoding rat glial cell line-derived neurotrophic factor protects nigral neurons in a progressive 6-hydroxydopamine-induced degeneration model of Parkinson's disease in rats. *Proceedings of the National Academy of Sciences, USA,* **94,** 14083–14088.

MANGIARINI, L., SATHASIVAM, K., SELLER, M., COZENS, B., HARPER, A., HETHERINGTON, C., LAWTON, M., TROTTIER, Y., LEHRACH, H., DAVIES, S.W. and BATES, G.P. (1996) Exon 1 of the HD gene with an expanded CAG repeat is sufficient to cause a progressive neurological phenotype in transgenic mice. *Cell,* **87,** 493–506.

MANSUY, I.M., MAYFORD, M., JACOB, B., KENDEL, E.R. and BACH, M.E. (1998) Restricted and regulated overexpression reveals calcineurin as a key component in the transition from short-term to long-term memory. *Cell,* **92,** 39–49.

MAYFORD, M., BACH, M.E., HUANG, Y.Y., WANG, L., HAWKINS, R.D. and KANDEL, E.R. (1996) Control of memory formation through regulated expression of a CaMKII transgene. *Science,* **274,** 1678–1683.

MIETTINEN, P.J., BERGER, J.E., MENESES, J., PHUNG, Y., PEDERSEN, R.A., WERB, Z. and DERYNK, R. (1995) Epithelial immaturity and multiorgan failure in mice lacking epidermal growth factor receptor. *Nature,* **376,** 337–341.

MILNER, R.J. and SUTCLIFFE, J.G. (1983) Gene expression in rat brain. *Nucleic Acids Research,* **11,** 5497–5520.

MONAGHAN, A.P., BOCK, D., GASS, P., SCHWANGER, A., WOLFER, D.P., LIPP, H.P. and SCHUTZ, G. (1997) Defective limbic system in mice lacking the *tailless* gene. *Nature,* **390,** 515–517.

MORATALLA, R., EIBOL, B., VALLEJO, M. and GRAYBIEL, A.M. (1996) Network-level changes in expression of inducible Fos-Jun proteins in the striatum during chronic cocaine treatment and withdrawal. *Neuron,* **17,** 147–156.

MORGAN, J.I. and CURRAN, T. (1991) Stimulus-transcription coupling in the nervous system: involvement of the inducible proto-oncogenes fos and jun. *Annual Review of Neuroscience,* **14,** 421–451.

NAIR, S.M., WERKMAN, T.R., CRAIG, J., FINNELL, R., JOELS, M. and EBERWINE, J.H. (1998) Corticosteroid regulation of ion channel conductances and mRNA levels in individual hippocampal CA1 neurons. *Journal of Neuroscience,* **18,** 2685–2696.

NELSON, R.J., DEMAS, G.E., HUANG, P.L., FISHMAN, M.C. DAWSON, V.L., DAWSON, T.M. and SNYDER, S.H. (1995) Behavioural abnormalities in male mice lacking neuronal nitric oxide synthase. *Nature,* **378,** 383–386.

NESTLER, E.J. and AGHAJANIAN, G.K. (1997) Molecular and cellular basis of addiction. *Science,* **278,** 58–63.

OKUBO, Y., SUHARA, T., SUZUKI, K., KOBAYASHI, K., SASSA, T., SUDO, Y., MATSUSHIMA, E., IYO, M., TATENO, Y. and TORU, M. (1997) Decreased prefrontal dopamine D1 receptors in schizophrenia revealed by PET. *Nature,* **385,** 634–636.

ORDWAY, J.M., TALLAKSEN-GREENE, S., GUTEKUNST, C.A., BERNSTEIN, E.M., CLEARLEY, J.A., WIENER, H.W., DURE, L.S., LINDSEY, R., HERSCH, S.M., JOPE, R.S., ALBIN, R.L. and DETLOFF, P.J. (1997) Ectopically expressed CAG repeats cause intranuclear inclusions and a progressive late onset neurological phenotype in the mouse. *Cell,* **91,** 753–763.

PELLEGRINI-GIAMPIETRO, D., GORTER, J.A., BENNETT, M.V.L. and ZUKIN, R.S. (1997) The GluR2 (GluR-B) hypothesis: Ca^{2+} permeable AMPA receptors in neurological disorders. *Trends in Neuroscience,* **20,** 464–469.

PEPIN, M.C., POTHIER, F. and BARDEN, N. (1992) Impaired Type II glucocorticoid-receptor function in mice bearing antisense RNA transgene. *Nature,* **355,** 725–728.

PICCIOLI, P., DI LUZIO, A., AMANN, R., SCHULIGOI, R., SURANI, M.A., DONNERER, J. and CATTANEO, A. (1995) Neuroantibodies: ectopic expression of a recombinant anti-substance P antibody in the central nervous system of mice. *Neuron,* **15,** 373–384.

PRICE, D.L., SISODIA, S.S. and BORCHELT, D.R. (1998) Alzheimer's disease – when and why? *Nature Genetics* **19,** 314–316.

PRUSINER, S.B. (1997) Prion diseases and the BSE crisis. *Science* **278,** 245–251.

REPPERT, S.M. and WEAVER, D.R. (1997) Forward genetic approach strikes gold: cloning of a mammalian *Clock* gene. *Cell,* **89,** 487–490.

ROGERS, D.C., FISHER, E.M., BROWN, S.D., PETERS, J., HUNTER, A.J. and MARTIN, J.E. (1997) Behavioral and functional analysis of mouse phenotype: SHIRPA, a proposed protocol for comprehensive phenotype assessment. *Mammalian Genome,* **8,** 711–71.

SAIKI, R.K., GELFAND, D.H., STOFFEL, S., SCHARF, S.J., HIGUCHI, R., HORN, G.T., MULLIS, K.B. and ERLICH, H.A. (1988) Primer-directed enzymatic amplification of DNA with a thermostable DNA polymerase. *Science,* **239,** 487.

SAMBROOK, J., FRITSCH, E.F. and MANIATIS, T. (1989) *Molecular Cloning* 2nd edn. Cold Spring Harbor Laboratory Press, New York.

SAUMAN, I. and REPPERT, S.M. (1996) Circadian clock neurons in the silkmoth *Antheraea pernyi*: novel mechanisms of period protein regulation. *Neuron,* **17,** 889–900.

SCHNIEKE, A., HARBERS, K. and JAENISCH, R. (1983) Embryonic lethal mutation in mice induced by retrovirus insertion into the alpha 1(I) collagen gene. *Nature,* **304,** 315–320.

SCHNIEKE, A.E., KIND, A.J., RITCHIE, W.A., MYCOCK, K., SCOTT, A.R., RITCHIE, M., WILMUT, I., COLMAN, A. and CAMPBELL, K.H. (1997) Human factor IX transgenic sheep produced by transfer of nuclei from transfected fetal fibroblasts. *Science,* **278,** 2130–2133.

SEGAL, R.A. and GREENBERG, M.E. (1996) Intracellular signalling pathways activated by neurotrophic factors. *Annual Review of Neuroscience,* **19,** 463–489.

SHERIN, J.E., SHIROMANI, P.J., MCCARLEY, R.W. and SAPER, C.B. (1996) Activation of ventrolateral preoptic neurons during sleep. *Science,* **271,** 216–219.

SIBILIA, M. and WAGNER, E.F. (1995) Strain-dependent epithelial defects in mice lacking the EGF receptor. *Science,* **269,** 234–238.

SIRUGO, G., PAKSTIS, A.J., KIDD, K.K., MATTHYSSE, S., LEVY, D.L., HOLZMAN, P.S., PARNA, S.J., MCINNIS, M., BRESCHEL, T. and ROSS, C.A. (1997) Detection of a large CTG/CAG trinucleotide repeat expansion in a Danish schizophrenia kindred. *American Journal of Medical Genetics,* **74,** 546–548.

SISODIA, S. (1998) Nuclear inclusions in glutamine repeat disorders: Are they pernicious, coincidental, or beneficial? *Cell,* **95,** 1–4.

SOMMER, B., KEINANEN, K., VERDOORN, T.A., WISDEN, W., BURNASHEV, N., HERB, A., KOHLER, M., TAKAGI, T., SAKMANN, B. and SEEBURG, P.H. (1990). Flip and flop: a cell-specific functional switch in glutamate-operated channels of the CNS. *Science*, **249**, 1580–1585.

SON, H., HAWKINS, R.D., MARTIN, K., KIEBLER, M., HUANG, P.L., FISHMAN, M.C. and KANDEL, E.R. (1996) Long-term potentiation is reduced in mice that are doubly mutant in endothelial and neuronal nitric oxide synthase. *Cell*, **87**, 1015–1023.

TAKAHASHI, J.S., PINTO, L.H. and VITATERNA, M.H. (1994) Forward and reverse genetic approaches to behaviour in the mouse. *Science*, **264**, 1724–1733.

TEI, H., OKAMURA, H., SHIGEYOSHI, Y., FUKUHARA, C., OZAWA, R., HIROSE, M. and SAKAKI, Y. (1997) Circadian oscillation of a mammalian homologue of the *Drosophila* period gene. *Nature*, **389**, 512–516.

THREADGILL, D.W., DLUGOSZ, A.A., HANSEN, L.A. *et al*., (1995) Targeted disruption of mouse EGF receptor: effect of genetic background on mutant phenotype. *Science*, **269**, 230–234.

TSIEN, J.Z., HUERTA, P.T. and TONEGAWA, S. (1996) The essential role of hippocampal CA1 NMDA receptor-dependent synaptic plasticity in spatial memory. *Proceedings of the National Academy of Sciences USA*, **87**, 1327–1338.

ULLRICH, B., USHKARYOV, Y.A. and SUDHOF, T.C. (1995) Cartography of neurexins: more than 1000 isoforms generated by alternative splicing and expressed in distinct subsets of neurons. *Neuron*, **14**, 497–507.

VALTIN, H., SCHROEDER, H.A., BERISCHKE, K. and SOKOL, H.W. (1962) Familial hypothalamic diabetes insipidus in rats. *Nature*, **196**, 1109–1110.

VAN LEEUWEN, F., DE KLEIJN, D.P.V., VAN DEN HURK, H., *et al*., (1998) Frameshift mutants of β amyloid precursor protein and ubiquitin-B in Alzheimer's and Down patients. *Science*, **279**, 242–247.

VENKATESH, V., SI-HOE, S-L., MURPHY, D. and BRENNER, S. (1997) Transgenic rats reveal functional conservation of regulatory controls between the Fugu isotocin and rat oxytocin genes. *Proceedings of the National Academy of Sciences USA*, **94**, 12462–12466.

WANG, Z.Q., OVITT, C., GRIGORIADIS, A.E., MOHLE, S.U., RUTHER, U., and WAGNER, E.F. (1992) Bone and haematopoietic defects in mice lacking c-fos. *Nature*, **360**, 741–745.

WELSH, D.K., LOGOTHETIS, D.E., MEISTER, M. and REPPERT, S.M. (1995) Individual neurons dissociated from rat suprachiasmatic nucleus express independently phased circadian firing rhythms. *Neuron*, **14**, 697–706.

WHITE, J.K., AUERBACH, W., DUYAO, M.P., VONSATTEL, J-P., GUSELLA, J.P., JOYNER, A.L. and MACDONALD, M.E. (1997) Huntingtin is required for neurogenesis and is not impaired by the Huntington's disease CAG expansion. *Nature Genetics,* **17,** 404–410.

WILMUT, I., SCHNIEKE, A.E., MCWHIR, J., KIND, A.J. and CAMPBELL, K.H.S. (1997) Viable offspring from fetal and adult mammalian cells. *Nature,* **385,** 810–813.

WINDER, D.G., MANSUY, I.M., OSMAN, M., MOALLEM, T.M. and KANDEL, E.R. (1998) Genetic and pharmacological evidence for a novel, intermediate phase of long-term potentiation surpressed by calcineurin. *Cell,* **92,** 25–37.

WOLLNICK, F., BRYSCH, W., UHLMANN, E., GILLARDON, F., BRAVO, R., ZIMMERMANN, M., SCHLINGENSIEPEN, K.H. and HERDEGEN, T. (1995) Block of c-Fos and JunB expression by antisense oligonucleotides inhibits light-induced phase shifts of the mammalian circadian clock. *European Journal of Neuroscience,* **7,** 388–393.

YAN, S.D., FU, J., SOTO, C., CHEN, X., ZHU, H., AL-MOHANNA, F., COLLISON, K., ZHU, A., STERN, E., SAIDO, T., TOHYAMA, M., OGAWA, S., ROHER, A. and STERN, D. (1997) An intracellular protein that binds amyloid-β peptide and mediates neurotoxicity in Alzheimer's disease. *Nature,* **389,** 689–695.

ZENG, Q., CARTER, D.A. and MURPHY, D. (1994) Cell-specific expression of a vasopressin transgene in rats. *Journal of Neuroendocrinology,* **6,** 469–477.

Index